# YOUR KNOWLEDGE HAS VALUE

# Optimization of Pulses and Pulse Sequences for NMR Spectroscopy

Stella Slad

**Bibliographic information published by the German National Library:**

The German National Library lists this publication in the National Bibliography; detailed bibliographic data are available on the Internet at http://dnb.dnb.de.

ISBN: 9783346714152
This book is also available as an ebook.

© GRIN Publishing GmbH
Nymphenburger Straße 86
80636 München

Print and binding: Books on Demand GmbH, Norderstedt, Germany
Printed on acid-free paper from responsible sources.

The present work has been carefully prepared. Nevertheless, authors and publishers do not incur liability for the correctness of information, notes, links and advice as well as any printing errors.

GRIN web shop: https://www.grin.com/document/1269029

KARLSRUHE INSTITUTE OF TECHNOLOGY
FACULTY OF CHEMISTRY AND BIOSCIENCES

MASTER THESIS

# Optimization of Pulses and Pulse Sequences for NMR Spectroscopy

Stella Slad

06.05.2019

# Acknowledgements

I would like to thank Prof. Dr. Burkhard Luy who made it possible to work on these projects and gave answers to some very important questions. I would also like to thank my supervisor, Dr. David Goodwin, for help with a lot of different issues. And I would like to thank to all members of AK Bulu for all the activities and their support.

Finally, personal acknowledgement is given to my parents, Nella and Davyd Brytanchuk, my husband, Damian Slad, and my son, Michael Slad, for their support. I hope that one day Michael will understand the main ideas of this work.

# Abstract

Pulse engineering plays an important role in high-resolution NMR spectroscopy because the performance of existing pulses depends on experimental parameters like bandwidth or magnetic field inhomonegeities. The GRAPE optimization algorithm [1] can be used to find the best pulse for a given set of parameters. This method has been used to design band-selective pulses [2–4], robust broadband excitation and inversion pulses [5–8] and various universal rotation pulses [9].

The first part of this work is an extension of the systematic studies on broadband pulses [7–9]. This time the GRAPE algorithm is used to design broadband 30° and 60° excitation pulses as well as universal rotation pulses with the same flip angles. Correlations between the best achievable quality factor and pulse duration have been measured for different bandwidths and degrees of rf-inhomogeneity tolerance. Minimum pulse durations for a given quality factor have been evaluated and compared to studies of 90° and 180° pulses. The obtained pulse shapes are similar to previously published point-to-point and universal rotation pulses optimized with this method.

The second part of this work is concerned with the design of ultra-broadband $^{19}$F-CMPG [10, 11] and $^{19}$F-PROJECT [12] pulse sequences that could be used for ligand-based binding studies. The best CPMG sequence was a combination of a BURBOP-90 pulse with a BURBOP-180 pulse. For PROJECT, the best results were achieved using the same 90° pulse and a pair of BIBOP pulses instead of a universal rotation pulse. Simulations showed that the PROJECT sequence performs significantly better than the CPMG sequence in the presence of fluorine-fluorine couplings.

# Contents

# List of Figures

## List of Tables

# 1 Introduction

## 1.1 NMR spectroscopy

Nuclear Magnetic Resonance (NMR) spectroscopy is one of the most important analytical techniques used in Chemistry, Biochemistry and Biology [13, 14]. Nuclear spins precess with their Larmor frequency in a static magnetic field ($B_0$). This frequency depends on the strength of $B_0$, the type of the nucleus (e. g. $^1$H, $^2$H and $^{13}$C) and its chemical environment. An NMR experiment consists of one or multiple radiofrequency (rf) pulses that induce transitions between different spin states. At the end of the pulse sequence, the resonance response of the nuclei is detected and Fourier-transformed into a spectrum. A more detailed theoretical description of NMR is provided in section 2.1.

Important information obtained from high-resolution NMR spectra includes chemical shifts and coupling constants. The chemical shift of a measured signal depends on the chemical groups neighbouring this nucleus. Coupling between spins is observed as signal splittings in the spectrum. Knowing the values of the coupling constants can help to determine the conformation of molecules and to distinguish between *cis*- and *trans*-isomers. Multi-dimensional correlation experiments provide additional information, e. g. which C atoms are bound to which H atoms.

The main area of application is structure elucidation of small molecules (e. g. products of a novel synthesis). This can be done efficiently with a set of one-dimentional and multi-dimensional correlation experiments [13]. In Biology and Biochemistry, NMR spectroscopy can be used for analysing complex mixtures of metabolites [15, 16], determining 3D structures of proteins [17] and investigating protein-ligand interactions [18, 19]. Other techniques like IR spectroscopy and mass spectrometry cannot provide as much information about the structure of molecules. Therefore, they can only be used for structure determination in conjunction with other techniques or if a large database, which includes the molecule of interest, is provided. X-ray crystallography is a very powerful technique for structure determination of proteins, but it requires growth of crystals. Finding out the right crystallization conditions can take a long time. In addition, the structure of a peptide/protein in solution can be different to the crystallized structure. Another advantage of NMR spectroscopy, compared to mass spectrometry and X-ray crystallography, is that the samples usually do not get damaged and can be reused [20].

## 1.2 Pulse optimization

Commonly used pulses have a rectangular shape and rotate spins by an angle $\alpha$ defined by the rf-amplitude and the pulse duration. These *hard pulses* perform well for small frequency bandwidths, e. g. in $^1$H spectroscopy, but must be calibrated before the experiment. In order to achieve uniform rotation over larger bandwidths, the rf-amplitude has to be increased. Some nuclei, for example $^{19}$F, have signals that would cover the 200-fold of a $^1$H spectrum. Rf-amplitudes cannot be increased indefinetely because very high rf-amplitudes could damage the sensitive electronics of the NMR spectrometer or the sample. If the rf-amplitude is significantly lower than the bandwidth or the pulse calibration was incorrect, signals at the edges of the spectrum have lower intensities. More complicated pulse sequences often require special pulses which are more robust with respect to large bandwidths and rf-amplitude variation.

A pulse with better performance was first introduced in 1979 by M. H. Levitt and R. Freeman [21]. They showed that a pulse composed of multiple hard pulses ( $90_y 180_x 90_y$) was more robust against $B_1$ inhomogeneity than one hard pulse. Later, more *composite pulses* were published and classified [22]. Since nowadays most rf-amplifiers are capable of continuously modulating pulse amplitude and phase, it is possible to use pulses with arbitrary shapes.

The GRadient Ascent Pulse Engineering (GRAPE) algorithm has been shown to be a very effective tool for pulse optimization [1]. This method has been used to design band-selective pulses [2–4], robust broadband excitation and inversion pulses [5–8] and universal rotation pulses [9]. In this work, GRAPE is used to design broadband 30° and 60° excitation pulses as well as universal rotation pulses and to explore their physical limits. In addition, it is used to design ultra-broadband universal rotation and inversion pulses for the CMPG [10, 11] and PROJECT [12] pulse sequences.

## 1.3 Exploring the physical limits of broadband 30° and 60° pulses

In some NMR experiments, pulses inducing flip angles smaller than 90° need to be used. A well-known example is the $\beta$-angle excitation routinely used in $^{13}$C-1D spectroscopy which relies on a relaxation-matched flip angle [23]. $\beta$-pulses are also used as mixing pulses in zz-spectroscopy [24–26], COSY experiments [27–32], and various heteronuclear correlation experiments [33–37].

In general, shaped pulses are not scalable with respect to the flip angle, but some pulse families can be scaled to arbitrary flip angles. They are referred to as rf-amplitude-dependent flip angle (RADFA) pulses [38, 39] and are very useful for determining the optimal flip angle for a given experiment. Currently, these pulses are the best performing pulses for experiments requiring small flip angle pulses.

For some experiments, the optimal flip angle is well-known, like in the case of $^{13}C$ $\beta$-angle excitation, where typically pulses with a flip angle of 30° are used. Here, optimizing broadband pulses with this specific flip angle may result in better performance compared to RADFA pulses. In this work, 30° and 60° pulses are optimized for broadband excitation and high tolerance to rf-variation. For some of the mentioned applications, universal rotation (UR) pulses are a better alternative to excitation pulses (more in detail in Section 2.3.1). Therefore, UR pulses are also optimized for 30° and 60° flip angles. One of the aims is to explore the physical limits of small flip angle pulses for a given set of parameters including pulse duration, maximum rf-amlitude and maximum rf-power.

## 1.4 Designing broadband universal rotation pulses for $^{19}F$ CPMG and $^{19}F$ PROJECT

$^{19}$F-NMR methods are gaining popularity in drug discovery programs because they yield simpler spectra than $^1$H-NMR experiments and are sensitive to weak protein-ligand interactions [40–43]. Binding of a small molecule to a protein can be detected easily with the Carr-Purcell-Meibom-Gill (CPMG) experiment [10, 11], as it leads to reduction of the ligand signals. However, evolution of fluorine-fluorine couplings during this experiment leads to anti-phase magnetization, observed as distorted signals within the spectrum. Therefore, when the conventional CPMG sequence is used with polyfluorinated compounds, spectra must be processed in magnitude mode resulting in a lower sensitivity.

In 2012, Aguilar et al. [12] developed a version of the CPMG sequence where homonuclear J coupling is refocused. This experiment is now referred to as PROJECT (Periodic Refocusing of J Evolution by Coherence Transfer) and the resulting spectra show no J modulation. This sequence has already been used for detecting binding of polyfluorinated small molecules to proteins [44]. Only signals within a bandwidth of 70 ppm were displayed in this study, but flourinated organic compounds cover a chemical shift range of approximately 300 ppm [18]. On a 400 MHz spectrometer this corresponds to a bandwidth of 113 kHz. Hard pulses, however, can only cover a bandwidth of approximately 20 - 30 kHz because the maximum rf-amplitude tolerated by the spectrometer probeheads is limited.

In this work, the GRAPE algorithm will be used to optimize ultra-broadband universal rotation and inversion pulses covering a bandwidth of 120 kHz. Different CPMG and PROJECT sequences will be compared to each other using simulations.

# 2 Theoretical background

Many textbooks on NMR spectroscopy contain information on Sections 2.1, 2.2 and 2.3, for example, "Understanding NMR Spectroscopy" by J. Keeler [45], "Ein- und zweidimensionale NMR-Spektroskopie" by H. Friebolin [13] and "Principles of Nuclear Magnetic Resonance in One and Two Dimensions" by R. R. Ernst [23]. If no source is specified in the text, the information originated from one of these books.

## 2.1 NMR spectroscopy

Three major principles are needed in order to understand one-dimensional NMR spectroscopy:

1. The behaviour of nuclear spins inside of a homogeneous static magnetic field (see Section 2.1.1)

2. The effect of short radiofrequency pulses on nuclear spins and the net magnetization in the sample (see Section 2.1.2) and

3. The behaviour of the net magnetization after the pulse (see Section 2.1.3).

An additional section will present the CPMG experiment and how it is used for detection of protein-ligand interactions.

### 2.1.1 Nuclear spins in a static magnetic field

In order to perform an NMR experiment, the sample is placed inside a strong static magnetic field $B_0$. This causes an energy difference $\Delta E$ between spin states with different magnetic spin quantum numbers $m_I$, the so-called Zeeman effect. The number of these states $n$ is defined by the nuclear spin quantum number $I$:

$$n = 2I + 1 \tag{1}$$

Most nuclei used in NMR have $I = 1/2$, hence, there are only two different states, corresponding to a parallel or antiparallel orientation of the spins with respect to $B_0$. The energy difference between the states depends on the strength of the static magnetic field and on the gyromagnetic ratio $\gamma$ of the nucleus (see Fig. 1.B).

$$\Delta E = \gamma \hbar B_0 \tag{2}$$

with $\hbar$ being the reduced Planck constant and $\gamma$ the gyromagnetic ratio of the nucleus. $\gamma$ is defined as the ratio between the nuclear angular momentum $P$ and the magnetic momentum $\mu$:

$$\gamma = \frac{\mu}{P} \tag{3}$$

In addition to $B_0$, there are local magnetic fields caused by surrounding electrons. These weaken the effect of $B_0$ on the neighboring spins, which is described by the shielding constant $\sigma$. The resulting magnetic field in the direction of $B_0$, defined as the z-axis, is now given by

$$B_z = B_0(1 - \sigma). \tag{4}$$

The spins precess around the $B_z$ axis at their Larmor frequencies $\omega_L$ (see Fig. 1.A).

$$\omega_L = |\gamma| B_z \tag{5}$$

The net magnetization $M$ is located along the z-axis and is proportional to the population difference of the two states. This is also referred to as equilibrium magnetization or polarization.

### 2.1.2 The effect of a short radiofrequency pulse

Only the x- and y-components of the net magnetization are detectable, therefore $M$ should be rotated into the xy-plane. In modern NMR spectrometers this task is accomplished by a radiofrequency pulse. The pulse generates a linearly oscillating magnetic field, which can be seen as a superposition of two components rotating in opposite directions around the z-axis. The component with the same sense of rotation as the precessing spins will be referred to as $B_1$. The effect of this magnetic field is understood best by changing the frame of reference to a frame rotating at the carrier frequency of the pulse $\omega_1$. In this frame, $B_1$ and nuclei rotating with $\omega_L = \nu_1$ are static and other spins precess around the z-axis. Their new frequencies are given by

$$\Omega = \omega_L - \omega_1 \tag{6}$$

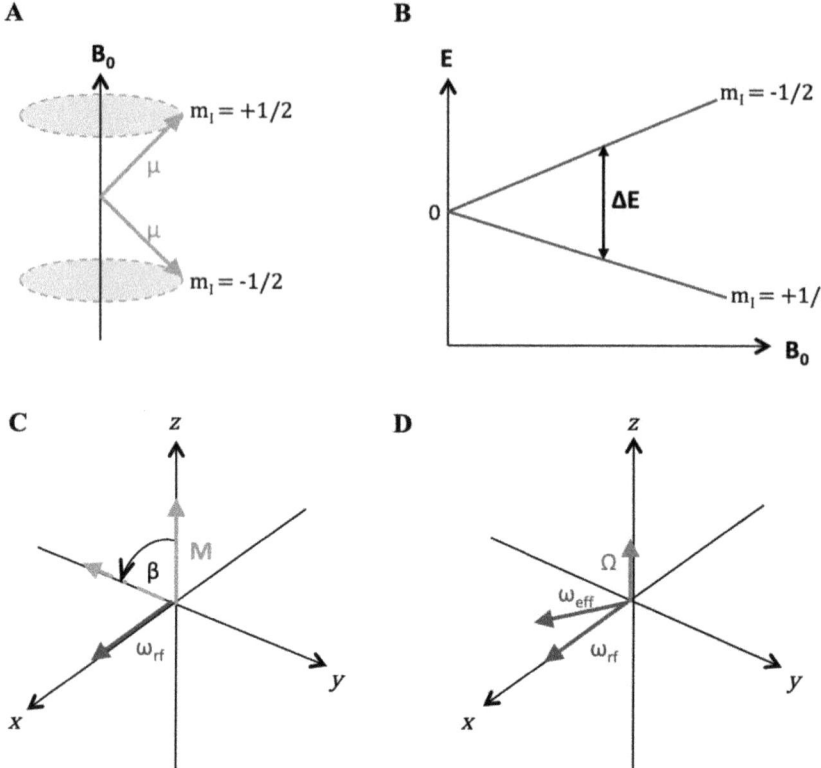

**Figure 1:** This is an illustration of the basic principles of an NMR experiment: In a static magnetic field, nuclear angular momenta $\mu$ precess around the $B_0$ axis (A). For a spin with $I = 1/2$, there are two possible spin states corresponding to two different orientations. The energy difference between these states depends on the strength of the static magnetic field (B). The easiest way to see the effect of a radiofrequency pulse on the equilibrium net magnetization $M$ is by changing the frame of reference to a frame rotating with the carrier frequency of the pulse. In this rotating frame, an on-resonance 90° pulse coming from the x-direction rotates the net magnetization about the x-axis resulting in -y-magnetization (C). If $\omega_L \neq \omega_1$, the rotation axis is defined by the effective field vector $\omega_{eff}$, the vector sum of the frequency offset $\Omega$ and the rf-amplitude $\omega_{rf}$ (D).

and are referred to as frequency offsets. In this work, they are often given in Hz corresponding to

$$\nu = \frac{\Omega}{2\pi} \tag{7}$$

Consider the simplest case, where all spins in the sample have the same Larmor frequency and $\Omega = 0$. Such a pulse is called an on-resonance pulse and rotates the net magnetization by an angle $\beta$ defined by the rf-amplitude $\omega_{rf}$ and the pulse duration $t_p$:

$$\beta = \omega_{rf} t_p \tag{8}$$

The rotation axis is aligned with the direction of $\omega_{rf}$, e. g. a $90°_x$ pulse rotates the magnetization by an angle of $90°$ about the x-axis (see Fig. 1.C).

In practice, a molecule contains several nuclei of the same type which have different Larmor frequencies. As a consequence, the pulse cannot be on-resonance with all of them. In general, the effective field vector $\omega_{eff}$ given as the vector sum of the frequency offset $\Omega$ and the rf-amplitude $\omega_{rf}$ defines the rotation axis (see Fig. 1.D). If $\omega_{rf} \gg |\Omega|$, the effect of a hard pulse is approximately the same as of an on-resonance pulse.

### 2.1.3 Relaxation and signal detection

After a radiofrequency pulse the net magnetization of the sample slowly returns to its equilibrium value. This process is called relaxation and is described by the following equations.

$$\frac{dM_x(t)}{dt} = \frac{-M_x(t)}{T_2} \tag{9}$$

$$\frac{dM_y(t)}{dt} = \frac{-M_y(t)}{T_2} \tag{10}$$

$$\frac{dM_z(t)}{dt} = \frac{-M_z(t) + M_z(t_0)}{T_1} \tag{11}$$

The first two equations describe the decay of transverse magnetization due to dephasing of the spins, which is characterized by the time constant $T_2$. In contrast, the increase of longitudinal magnetization is determined by a different time constant, $T_1$. The main focus of this section will lie on $T_2$ relaxation because this is particularly relevant for understanding the CPMG experiment.

In general, relaxation is a result of many state transitions of individual spins that happen due to interactions with local magnetic fields oscillating at the Larmor frequency. In $^1$H NMR, the main mechanism contributing to transverse relaxation is the dipole-dipole coupling (DD). This is a through-space-interaction with the magnetic fields of the nearest magnetic nuclei. Due to molecular tumbling, the nucleus will experience an oscillating magnetic field.

Another source for fluctuating fields are the surrounding electrons: If the distribution of electrons around a nucleus is not symmetrical, then their shielding depends on the orientation of the molecule. This phenomenon is referred to as chemical shift anisotropy (CSA) and is the main $T_2$ mechanism in $^{19}$F spectroscopy.

Rotational diffusion of molecules is described by the correlation time constant $\tau_c$, the average time that it takes a molecule to rotate by one radian. Assuming a spherical molecule, $\tau_c$ is proportional to the viscosity $\eta$ of the solvent and the third power of the hydrodynamic radius $r_H$ of the molecule:

$$\tau_c = \frac{4\pi\eta r_H^3}{3k_B T},\tag{12}$$

where $k_B$ is the Boltzmann constant and $T$ the sample temperature. As $r_H^3$ is proportional to the molecular weight (MW), rotational correlation times of proteins are significantly longer than of small molecules. For example, a ligand with $MW = 300$ Da would have $\tau_c$ of about 0.2 ns in aqueous solution. The same molecule will rotate with $\tau_c \approx 20$ ns, if it is bound to a 30 kDa protein. Changes in $\tau_c$ times can be observed by measuring $T_2$ relaxation times: Protein-bound ligands have significantly shorter $T_2$ times than free ligands.

As mentioned in the previous section, only transverse magnetization can be detected in NMR experiments. The precessing net magnetization of the sample induces a current in an rf-coil located around the sample. This time-dependent signal is referred to as the Free Induction Decay (FID) and is Fourier-transformed into a frequency-spectrum. The measured decay of transverse magnetization after a pulse is related to $T_2$, but it is also influenced by the inhomogeneity of the $B_0$ field:

$$S(t) = S(0)\exp\{i\Omega t\}\exp\left\{\frac{-t}{T_2^*}\right\},\tag{13}$$

$$\frac{1}{T_2^*} = \frac{1}{T_2} + \frac{1}{T_2^\dagger}\tag{14}$$

$S(0)$ denotes the amplitude at the beginning of the FID, $\Omega$ the frequency offset of the spin, $T_2^*$ the time constant of the signal decay and $T_2^\dagger$ the time constant of the inhomogeneous decay.

## 2.2  The CPMG experiment

The inhomogeneous decay is caused by different $B_0$ strengths in different parts of the sample, which means that the same nuclei have different frequency offsets $\Omega$. This offset-evolution can be refocused by a spin-echo element: a variable delay $\Delta$ followed by a $180°_y$ pulse and another delay of the same duration. If the spin-echo is applied after an excitation pulse (see Fig. 2), the detected signal is not influenced by the inhomogeneous decay. On the other hand, the signal reduction due to $T_2$ relaxation is an irreversible process and will be observed.

The sequence introduced here is the Carr-Purcell-Meiboom-Gill (CPMG) [10, 11] experiment. Multiple 1D-spectra with different numbers of spin-echoes $n$ are recorded and the $T_2$ times are obtained by fitting the following function to the measured signal intensities

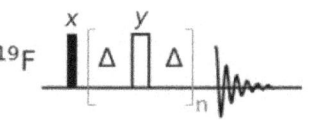

Figure 2: The basic CPMG sequence consists of a 90 ° pulse, a free precession period $2\Delta$ and a $180°_y$ pulse in the center of this time period. At the end of the sequence, all magnetization is aligned along the same axis, independent of $\Delta$ or the size of the offset $\Omega$.

$$f(\Delta) = I_0 \exp\left\{\frac{-2\Delta}{T_2}\right\} \qquad (15)$$

with $I_0$ being the maximum signal intensity for $\Delta = 0$. In order to detect ligand binding, the CPMG experiment has to be performed in the presence and in the absense of the protein [18, 19]. Setting $\Delta$ to $100 - 400$ ms removes the fast relaxing signals from the spectrum so that the relaxation-edited spectrum recorded in the presence of protein does not show signals from bound ligands. A subtraction of the two spectra results in a spectrum showing only signal from bound ligands.

[19]F-based methods are particularly useful for determining weak binding ligands because the main $T_2$ relaxation mechanisms, CSA, is particularly sensitive to changes in the rotational correlation time. However, currently available CPMG sequences cannot be applied to the whole [19]F bandwidth. As a results, mixtures for studies with multiple ligands have to be designed carefully so that the signals only cover a part of the spectrum (typically 40 ppm). Therefore, an ultra-broadband CPMG sequence would allow a higher throughput.

## 2.3 Quantum mechanical description of nuclear spins

The behaviour of spins during an NMR experiment is often illustrated using net magnetization vectors. However, a quantum mechanical approach is necessary for an exact description of spin-spin interactions.

### 2.3.1 Basic concepts

In quantum mechanics, all measurable quantities of a system are represented by operators, which will be denoted as bold letters in this thesis. For instance, the Hamilton operator $\mathbf{H}$ corresponds to the total energy of a system. The time-dependent Schrödinger equation is a differential equation that describes the time-evolution of the state $|\psi(t)\rangle$ of a quantum system:

$$\frac{\partial}{\partial t}|\psi(t)\rangle = -i\mathbf{H}|\psi(t)\rangle, \tag{16}$$

where In NMR spectroscopy, it is common to measure energy in angular frequency units, therefore the usual factor of $\hbar$ is not included in $\mathbf{H}$. The state $|\psi(t)\rangle$ can be expressed as a linear combination of the $n$ eigenfunctions $|\psi_i(t)\rangle$ of the Hamiltonian

$$|\psi(t)\rangle = \sum_{i=1}^{n} c_i(t)|\psi_i\rangle, \tag{17}$$

with time-dependent coefficients $c_i(t)$.

For a nucleus with spin quantum number $I = 1/2$, there are only two eigenfunctions corresponding to the two values of the magnetic quantum number $m_I = \pm 1/2$. They are commonly denoted as $|\alpha\rangle$ and $|\beta\rangle$.

$$|\psi\rangle = c_\alpha |\alpha\rangle + c_\beta |\beta\rangle \tag{18}$$

### 2.3.2 The density matrix

When dealing with multiple spins, it is important that the statistical information is evaluated correctly. A commonly used method to describe the state of an ensemble of spins is the density matrix formalism. The density matrix $\rho_i$ of a pure state $|\psi_i\rangle$ is defined as follows:

$$\rho_i = |\psi_i\rangle \langle\psi_i| \tag{19}$$

11

The density operator of a mixed state is obtained as a sum of $\rho_i$ multiplied by the probability $P_j$ that the system is in the pure state $|\psi_i\rangle$:

$$\rho_i = \sum_j P_j \, |\psi_i\rangle \langle \psi_i| \tag{20}$$

The density matrix can also be expressed as a linear combination of the angular momentum operators:

$$\rho = a_x(t)\mathbf{I_x} + a_y(t)\mathbf{I_y} + a_z(t)\mathbf{I_z} \tag{21}$$

The bulk magnetization components $M_x$, $M_y$ and $M_z$ are directly proportional to the coefficients $a_i$. Therefore, the expectation value of an observable A can be easily calculated with

$$\langle \mathbf{A} \rangle = \text{tr}\{\rho \, \mathbf{A}\} \tag{22}$$

### 2.3.3 The Hamiltonian

The Hamiltonian of a system with $N$ spins is a sum of terms resulting from different interactions:

$$\mathbf{H} = \mathbf{H_Z} + \mathbf{H_J} + \mathbf{H_{rf}} \tag{23}$$

The contribution from the magnetic field $B_z$, called the Zeeman-term, is given as

$$\mathbf{H_Z} = -\sum_{j=1}^{N} \gamma_j \mathbf{I}_{j,z} B_z \tag{24}$$

The main interaction between two spins in solution NMR spectroscopy is the scalar coupling, which is an electron-mediated interaction. In general, it has the form

$$\mathbf{H_J} = \sum_{i<j}^{N} 2\pi J_{ij}(\mathbf{I}_{i,x}\mathbf{I}_{j,x} + \mathbf{I}_{i,y}\mathbf{I}_{j,y} + \mathbf{I}_{i,z}\mathbf{I}_{j,z}), \tag{25}$$

where $J_{ij}$ denotes the coupling constant. For weak coupling, i. e. if $2\pi|J_{ij}| \ll |\omega_{L,i} - \omega_{L,j}|$, the xx- and yy-terms can be neglected resulting in

$$\mathbf{H_{J,weak}} = \sum_{i<j}^{N} 2\pi J_{ij} \mathbf{I}_{i,z}\mathbf{I}_{j,z} \tag{26}$$

The oscillating magnetic field produced by a radio-frequency pulse corresponds to a time-dependent Hamiltonian. However, in the rotating frame this interaction is given as

$$\mathbf{H}_{\mathrm{rf}} = -B_1 \sum_{j=1}^{N} \gamma_i (\mathbf{I}_{j,\mathrm{x}} \cos\phi + \mathbf{I}_{j,\mathrm{y}} \sin\phi) \tag{27}$$

### 2.3.4 The Liouville-von-Neumann equation

The equation of motion of the spin density matrix is the Liouville-von-Neumann equation:

$$\frac{\partial \rho}{\partial t} = -i\hbar [\mathbf{H}, \rho] \tag{28}$$

If the Hamiltonian $\mathbf{H}$ is constant, this differential equation has an analytical solution with respect to the time propagation:

$$\rho(t) = \mathbf{U} \rho_0 \mathbf{U}^\dagger \tag{29}$$

with the time propagator $\mathbf{U} = \exp\{-it\mathbf{H}\}$.

## 2.4 Gradient ascent pulse engineering

### 2.4.1 Classes of NMR pulses

There are two major types of pulses that should be discriminated in pulse engineering: *point-to-point* (PP) and *universal rotation* (UR) pulses. A pulse of the first type can only transfer one initial magnetization component to a specific target component. For example, z-magnetization can be transformed to x-magnetization. However, the final orientation of vectors that are orthogonal to z can be anywhere on the yz-plane. In contrast, UR pulses rotate an initial magnetization vector by a defined rotation angle and axis. Rectangular pulses would act as UR pulses, if they were indefinitely short. Given the technological limitations of the maximum rf-amplitude, they only act like this for a relatively small bandwidth.

### 2.4.2 Optimal control theory

*Optimal Control Theory* (OCT) is a mathematical field that is concerned with maximizing the performance of a controlled process. The forces on the system are devided into two groups,

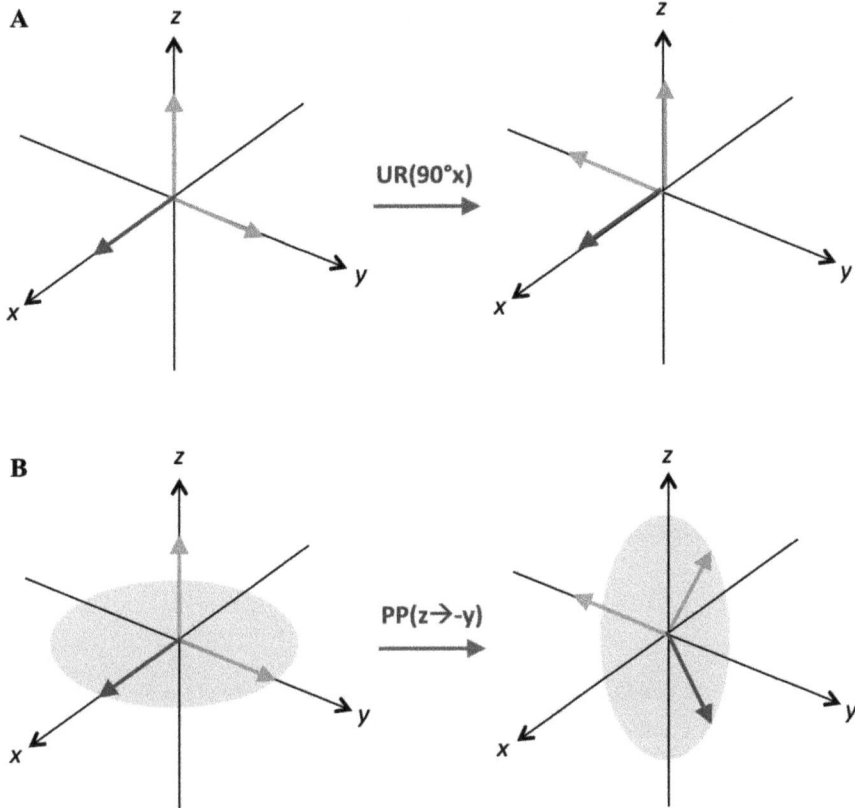

**Figure 3:** A universal rotation pulse (A) rotates all components of the initial magnetization vector ($M_x$: purple, $M_y$: orange and $M_z$: green) by a defined rotation angle and axis, for example, a $90°_x$ pulse. A point-to-point pulse only transfers one initial magnetization component to a specified target component, for example, $z \rightarrow -y$ (B). After this pulse, the vectors that were orthogonal to $M_z$ will be somewhere in the z-y-plane, their orientation is not taken care of.

controllable inputs (=controls) $u$ and disturbances $w$. Under their influence the system changes its state. The mathematical problem is finding the optimal controls. For this purpose, it is necessary to specify a cost function or a quality factor of the process that has to be minimized or maximized, respectively [46].

In NMR spectroscopy, the controls are the x- and y-components of the $B_1$ field and the disturbances are the frequency offsets $\nu$ of the spins. A pulse consisting of $n$ rectangular segments can be described using the control vector:

$$
\vec{u} = \begin{pmatrix} u_x(1) \\ u_y(1) \\ u_x(2) \\ u_y(2) \\ \vdots \\ u_x(n) \\ u_y(n) \end{pmatrix}
\tag{30}
$$

The cost function depends on the specific problem. In this work, two types of cost functions will be used:

1. The cost function for PP pulses is calculated as the overlap between the desired target state $C$ and the final state in the end of the pulse $\rho(t_p)$:

$$
\Phi_{PP} = \langle C|\rho(t_p)\rangle = \mathrm{tr}\{C^\dagger \rho(t_p)\}
\tag{31}
$$

2. The cost function for UR pulses is calculated as the overlap between the desired target propagator $\mathbf{U}_F$ and the propagator in the end of the pulse $\mathbf{U}_{\mathrm{eff}}$:

$$
\Phi_{UR} = \langle \mathbf{U}_F | \mathbf{U}_{\mathrm{eff}}\rangle
\tag{32}
$$

The gradient-based optimization algorithm will be explained in detail in the next section.

### 2.4.3  Optimization of point-to-point pulses

The GRadient Ascent Pulse Engineering (GRAPE) algorithm was first used for the design of broadband excitation and inversion pulses [7,8]. These are point-to-point problems: A defined inital state $\rho(t_0) = \rho_0$ should be transferred into a defined final state $C$ most efficiently. The cost function can be defined as

$$
\Phi_{PP} = \langle C|\rho(t_p)\rangle
\tag{33}
$$

The effect of a shaped pulse on a spin system can be described best using the piecewise-constant approximation, which actually resembles the experimental implementation of such pulses in modern NMR spectrometers. The pulse shape can be devided into $n$ slices of duration $\Delta t$ with constant control amplitudes $u_x(j)$ and $u_y(j)$. Then the time-evolution of the spin-system during a time step $j$ is given by the propagator

$$\mathbf{U}_j = \exp\{-i\Delta t(\mathbf{H_0} + u_x(j)\mathbf{H_x} + u_y(j)\mathbf{H_y})\} \tag{34}$$

$\mathbf{H_0}$ denotes the drift Hamiltonian, which includes the chemical shift and coupling terms. $\mathbf{H_x}$ and $\mathbf{H_y}$ are the control Hamiltonians. If the values of the controls are known, the density operator in the end of the pulse $\rho(t_p) = \rho_n$ can be calculated from the initial state $\rho_0$:

$$\rho_n = \mathbf{U}_n...\mathbf{U}_1\rho_0\mathbf{U}_1^\dagger...\mathbf{U}_n^\dagger \tag{35}$$

The desired final state $C = \lambda_n$ can also be back-propagated using the same formalism:

$$\lambda_0 = \mathbf{U}_1^\dagger...\mathbf{U}_n^\dagger\lambda_n\mathbf{U}_n...\mathbf{U}_1 \tag{36}$$

$\rho(t_p)$ in Eq. 33 can be replaced by the expression from Eq. 36:

$$\Phi_{PP} = \left\langle C \middle| \mathbf{U}_n...\mathbf{U}_1\rho_0\mathbf{U}_1^\dagger...\mathbf{U}_n^\dagger \right\rangle \tag{37}$$

This can be rearranged showing that $\Phi_{PP}$ equals the overlap of the trajectories of $\rho$ and $C$ at an arbitrary point in time $j\Delta t$:

$$\Phi_{PP} = \left\langle \underbrace{\mathbf{U}_{j+1}^\dagger...\mathbf{U}_n^\dagger\lambda_n\mathbf{U}_n...\mathbf{U}_{j+1}}_{\lambda_j} \middle| \underbrace{\mathbf{U}_j...\mathbf{U}_1\rho_0\mathbf{U}_1^\dagger...\mathbf{U}_j^\dagger}_{\rho_j} \right\rangle \tag{38}$$

Thus, the optimal pulse would result in a maximum overlap between the two trajectories (see dashed line in Fig. 4). The first order gradient of the cost function with respect to $u_k(j)$ can be calculated as:

$$\frac{\delta\Phi_{PP}}{\delta u_k(j)} \approx -\left\langle \lambda_j \middle| i\Delta t[\mathbf{H}_k, \rho_j] \right\rangle, \tag{39}$$

with $k \in \{x, y\}$. In order to maximize $\Phi_{PP}$, the controls $u_k(j)$ have to be modified according to

$$u_k(j) \rightarrow u_k(j) + \epsilon\frac{\delta\Phi_{PP}}{\delta u_k(j)}, \tag{40}$$

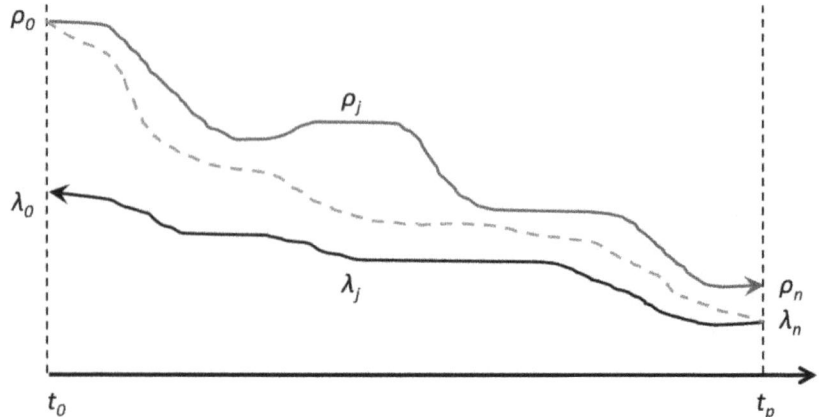

**Figure 4:** Trajectories of the initial state $\rho_0$ (dark blue) and the back-propagated desired target state $\lambda_n$ (black) during the pulse. The optimal trajectory maximizes the overlap between the two trajectories (dashed blue line).

where $\epsilon$ is a small step size.

In order to design of broadband PP pulses with $B_1$ inhomogeneity compensation it is necessary to calculate the cost function and the gradient for $n_{\text{off}}$ offsets linearly distributed over the desired bandwdith $\Delta\nu$ and $n_{\text{rf}}$ different rf-amplitudes $\nu_{\text{rf}}$. The global quality factor and the global gradient can be calculated as averages.

$$\overline{\Phi_{\text{PP}}} = \frac{1}{n_{\text{off}}n_{\text{rf}}} \sum_{i=1}^{n_{\text{off}}} \sum_{l=1}^{n_{\text{rf}}} \Phi_{\text{PP}}(\nu_{\text{off}}^i, \nu_{\text{rf}}^l) \tag{41}$$

$$\overline{\Gamma_k(j)} = \frac{1}{n_{\text{off}}n_{\text{rf}}} \sum_{i=1}^{n_{\text{off}}} \sum_{l=1}^{n_{\text{rf}}} \Gamma_k(j, \nu_{\text{off}}^i, \nu_{\text{rf}}^l) \tag{42}$$

**GRAPE algorithm for point-to-point pulses**

1. Guess initial controls $u_k(j)$

2. Starting from $\rho_0$, calculate $\rho_n = \mathbf{U}_j...\mathbf{U}_1\rho_0\mathbf{U}_1^\dagger...\mathbf{U}_j^\dagger$ for all $j \leq n$.

3. Starting from $\lambda_n = C$, calculate $\lambda_j = \mathbf{U}_{j+1}^\dagger...\mathbf{U}_n^\dagger C\mathbf{U}_n...\mathbf{U}_{j+1}$ for all $j \leq n$.

4. Evaluate individual local gradients $\frac{\delta\Phi_{\text{UR}}}{\delta u_k(j)}$ according to 39.

5. Repeat steps 2-4 for all offsets and all rf-amplitudes. Then calculate the gradient of the global quality factor $\overline{\Gamma_{\text{PP}}}$ and update the $2 \times n$ control amplitudes according to $u_k(j) \rightarrow u_k(j) + \epsilon\overline{\Gamma_k(j)}$.

6. Restrict controls to a maximum rf-amplitude or enforce any other restrictions

7. With these new controls, go to step 2 until a chosen convergence criterion is satisfied.

Here, the algorithm is described using spin density matrices $\rho$, but in the case of uncoupled spins, it is possible to use magnetization vectors and Bloch equations.

$$\frac{\mathrm{d}}{\mathrm{d}t}\overrightarrow{M}(t) = \overrightarrow{\omega_{\text{eff}}} \times \overrightarrow{M}(t) \tag{43}$$

The Bloch equations can be written in matrix representation including relaxation terms.

$$\frac{\mathrm{d}}{\mathrm{d}t}\overrightarrow{M}(t) = \begin{pmatrix} -\frac{1}{T_2} & \gamma B_z & -\gamma B_y \\ -\gamma B_z & -\frac{1}{T_2} & \gamma B_x \\ \gamma B_y & -\gamma B_x & -\frac{1}{T_1} \end{pmatrix} (\overrightarrow{M}(t) - \overrightarrow{M}_0) \tag{44}$$

This approximation can be used for the design of excitation and inversion pulses and speeds up the calculations significantly.

### 2.4.4 Optimization of universal rotation pulses

The GRAPE algorithm can be modified to optimize universal rotations as described in [1, 9]. Instead of defining the initial and the target state, we have to define the rotation axis and the flip angle. The local quality factor can be defined as

$$\Phi_{\text{UR}} = \frac{1}{2}\langle\mathbf{U}_F|\mathbf{U}_{\text{eff}}\rangle \tag{45}$$

Then the gradient can be calculated using the first order approximation:

$$\frac{\delta \Phi_{UR}}{\delta u_k(j)} \approx -\frac{1}{2} Re \, \langle P_j | i\Delta t \mathbf{H}_k X_j \rangle \tag{46}$$

with $X_j = \mathbf{U}_j...\mathbf{U}_1$ and $P_j = \mathbf{U}_{j+1}^\dagger...\mathbf{U}_n^\dagger \mathbf{U}_F$.

For the design of broadband UR pulses with $B_1$ inhomogeneity compensation, the global quality factor $\overline{\Phi_{UR}}$ and its gradient $\overline{\Gamma_{UR}}$ are calculated as ensemble averages, like in the case of PP pulses.

**GRAPE algorithm for UR pulses:**

1. Guess initial controls $u_k(j)$

2. Starting from $\mathbf{U}_0 = \mathbb{1}$, calculate $X_j = \mathbf{U}_j \mathbf{U}_{j-1}...\mathbf{U}_1\mathbf{U}_0$ for all $j \leq n$.

3. Starting from the desired propagator $\mathbf{U}_F$, calculate $P_j$ for all $j \leq n$.

4. Evaluate individual local gradients $\frac{\delta \Phi_{UR}}{\delta u_k(j)}$ according to Eq. (46).

5. Repeat steps 2-4 for all offsets and all rf-amplitudes. Then calculate the gradient of the global quality factor $\overline{\Gamma_{UR}}$ and update the $2 \times n$ control amplitudes according to $u_k(j) \rightarrow u_k(j) + \epsilon \overline{\Gamma_k(j)}$.

6. Restrict controls to a maximum rf-amplitude or enforce any other restrictions

7. With these new controls, go to step 2 until convergence has been reached

## 2.5 Broadband pulses in NMR spectroscopy

### 2.5.1 Adiabatic pulses

An important class of broadband pulses with a high degree of tolerance to rf-amplitude inhomogeneity are the *Adiabatic pulses* [47, 48]. This pulse type is inspired by the early NMR experiments where the amplitude of the static magnetic field $B_0$ was swept in the presence of continuous wave rf-irradiation [49, 50]. Nowadays, it is easier to keep $B_0$ constant and to sweep the pulse frequency $\omega_1(t)$. These adiabatic rotations have to be much faster than the relaxation times $T_1$ and $T_2$, thus, this technique is also referred to as adiabatic rapid passage [51].

Compared to hard pulses, adiabatic pulses are known to cover much broader bandwidths while minimizing rf-power dissipation.

Most commonly used adiabatic pulses are hyperbolic secant pulses [52], CHIRP [53] and WURST pulses [54, 55]. Applications of adiabatic pulses include solvent suppression [56–59], spectral editing based on spin-spin-coupling [60, 61] and polarization transfer [62–64]. However, most adiabatic pulses can only perform one or two of these transformations:

1. Excitation: Rotating longitudinal Magnetization $M_z$ onto the transverse plane. This type of frequency sweep is known as *adiabatic half-passage* (AHP).

2. Inversion: $M_z \rightarrow -M_z$. This process is referred to as *adiabatic full-passage* (AFP).

The best way to describe an adiabatic pulse is by using a frequency-modulated frame of the rf-frequency, the $\omega_1$ frame (x',y',z'). The effective magnetic field $\omega_{eff}$ for a spin with Larmor frequency $\omega_L$ is a vector sum of two components, as illustrated in Fig. 5.

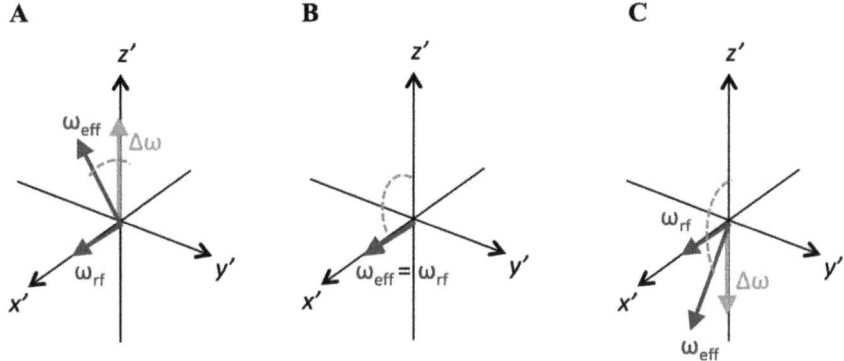

**Figure 5:** Adiabatic passage in a frequency modulated frame: The effective field vector $\omega_{eff}$ (blue) and its field components $\omega_1$ (purple) and $\Delta\omega$ (green) in the $\omega_{rf}$ frame of reference. The dashed line corresponds to the angle alpha, by which the effective field vector has been rotated during the sweep. In A, the rf-frequency $\omega_{rf}$ is much lower than the Larmor frequency - thus, the effective field vector is almost collinear with the z'-axis. In B, the pulse is on-resonant: $\omega_{rf} = \omega_L$. Then, $\Delta\omega = 0$, and $\omega_{eff}$ is aligned with the x'-axis. In C, $\omega_{rf} \gg \omega_L$ and $\omega_{eff}$ is oriented along z'. This figure was inspired by [48].

In the beginning of the adiabatic passage, the pulse frequency is far below resonance for all spins ($\omega_1 \ll \omega_L$). This means that $\Delta\omega \gg \omega_{rf}$ and therefore $\omega_{eff} \approx \Delta\omega$. Thus, the initial orientation of $\omega_{eff}$ is approximately collinear with z'. During the frequency sweep, $\omega_{eff}$ rotates towards the transverse plane with the instantaneous angular velocity $\dfrac{d\alpha}{dt}$. To perform an AHP, $\omega_{\rm rf}$ is increased, until resonance is reached, thus, $\omega_{\rm eff} = \omega_{\rm rf}$. This happens at different times for spins with different Larmor frequencies. For an AFP, the frequency sweep continues until $\omega_{\rm eff}$ is oriented along z' for all isochromats.

The bulk magnetization vector follows $\omega_{\rm eff}$ if the following condition is fulfilled for all times during the pulse:

$$|\omega_{\rm eff}(t)| \gg \left|\frac{d\alpha}{dt}\right| \tag{47}$$

This inequality is well-known as the *adiabatic condition* and is the main guiding principle in the design and optimization of frequency-modulation functions.

### 2.5.2 BIR-4 pulses

Some adiabatic pulses can perform uniform rotations with a high insensitivity to $B_1$ inhomogeneity - they are known as BIR pulses [65, 66]. The best performing pulse, BIR-4, is a composite pulse, consisting of four segments, each of which is either an adiabatic half-passage or a time-reversed adiabatic half-passage (see Fig. 6). This pulse shape has been derived using a symmetrized recursive expansion procedure. The amplitudes and phases are defined as follows.

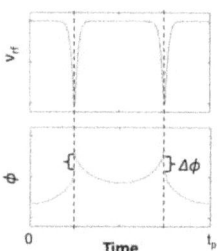

Figure 6: Typical BIR-4 pulse shape. The dashed lines indicate the phase shifts between segments.

Segment 1 ($0 < t \leq 0.25t_{\rm p}$):

$$\omega_{\rm rf}(t) = \omega_{\rm rf,max}\tan\{\zeta(1 - 4t/t_{\rm p}\} \tag{48}$$

$$\phi(t) = \frac{\phi_{\rm max}}{4} - \frac{\Delta\omega t_{\rm p}}{4\kappa\tan\kappa}\ln\left\{\frac{\cos(4\kappa t/t_{\rm p})}{\cos\kappa}\right\} \tag{49}$$

Segment 2 ($0.25t_{\rm p} < t \leq 0.5t_{\rm p}$):

$$\omega_{\rm rf}(t) = \omega_{\rm rf}(0.5t_{\rm p} - t) \tag{50}$$

$$\phi(t) = \phi(0.5t_{\rm p} - t) + \Delta\phi_1 \tag{51}$$

21

Segment 3 $(0.5t_p < t \leq 0.75t_p)$:

$$\omega_{\mathrm{rf}}(t) = \omega_{\mathrm{rf}}(t - 0.5t_p) \tag{52}$$

$$\phi(t) = \phi(t - 0.5t_p) + \Delta\phi_1 \tag{53}$$

Segment 4 $(0.75t_p < t \leq t_p)$:

$$\omega_{\mathrm{rf}}(t) = \omega_{\mathrm{rf}}(t_p - t) \tag{54}$$

$$\phi(t) = \phi(t_p - t) \tag{55}$$

$\omega_{\mathrm{rf,max}}$ denotes the maximum rf-amplitude, $\Delta\omega$ the range of the frequency sweep and $t_p$ the pulse duration. $\zeta$ and $\kappa$ are pulse shape parameters. The phase shift $\Delta\phi_1$ between segments 1 and 2 determines the net rotation angle:

$$\beta = 2(\Delta\phi_1 - \pi) \tag{56}$$

### 2.5.3 CHIRP pulses

Broadband excitation or inversion can also be accomplished by a linear frequency sweep. These pulses are called linear CHIRP pulses [53] and can be described by the following formula:

$$f(t) = f_0 + at \tag{57}$$

where $f(t)$ is the frequency of the pulse and $f_0$ is the frequency at the beginning of the pulse and a is the sweep rate. The sweep rate is defined by the desired offset range and pulse duration

$$a = \frac{\Delta\nu}{t_p} \tag{58}$$

If the rf-amplitude $\nu_{\mathrm{rf}}$ is fix, then, according to [67], the optimal sweep rate equals

$$a = \frac{\nu_{\mathrm{rf}}^2}{0.0762} \tag{59}$$

Traditionally, CHIRP pulses have constant rf-amplitude, however, pulses with smoothed edges act like adiabatic pulses and have cleaner inversion/excitation profiles [67].

The CHIRP pulses can cover up to 50 times larger bandwidths than their rf-amplitude and they are very robust against rf-inhomogeneity. However, for low rf-amplitudes and large bandwidths, the pulses tend to be very long ($> 1$ ms). This can lead to problems in systems with short longitudinal relaxation times $T_2$.

### 2.5.4 Broadband excitation and inversion pulses (BEBOP and BIBOP)

The broadband excitation and inversion pulses (BEBOP and BIBOP) are pulses that have been optimized using the GRAPE algorithm for PP pulses with limited rf-amplitude [7]. In contrast, power-BEBOP and power-BIBOP pulses [8] have limited average rf-amplitude, which corresponds to limiting the pulse power during optimization. With this approach, the peak rf-amplitude can be much higher than the given limit for average rf-amplitude. This leads to higher quality factors, especially if the bandwidth is larger than the maximum allowed rf-amplitude for BEBOP/BIBOP pulses. In order to achieve same quality factors as the rf-amplitude limited pulses, the rf-power limited pulses can be significantly shorter. For example, for $\Phi = 0.98$ and a bandwidth of $40 \, \mathrm{kHz}$ the minimum duration of corresponding power-BEBOP pulses is $80 \, \mu s$ - approximately half of the minimum pulse duration of BEBOP pulses.

### 2.5.5 Broadband universal rotation pulses (BURBOP)

As discussed before, for some applications in NMR, it is necessary to rotate all magnetization components by a defined axis and angle. For this purpose, a new family of universal rotation pulses has been optimized using the GRAPE algorithm. The resulting pulses were called Broadband Universal Rotations By Optimized Pulses (BURBOP) [9] and showed exceptional performance compared to previously published UR pulses.

These optimizations have shown that the convergence of the GRAPE algorithm depends on the definition of the cost function. Two different cost functions have been proposed and tested:

$$\Phi_1 = \frac{1}{4} \langle \mathbf{U_F} | \mathbf{U_{eff}} \rangle^2 \tag{60}$$

This cost function does not distinguish between $\mathbf{U_F}$ and $-\mathbf{U_F}$, corresponding to a 90° and a -270° rotation for 90° pulses. In general, the sense of rotation is described by a phase factor $e^{i\phi}$

that can take values of $-1$ or $1$ for $I = 1/2$.

$$\Phi_0 = \frac{1}{2} \langle \mathbf{U}_F | \mathbf{U}_{eff} \rangle \tag{61}$$

This cost function will only optimize $\mathbf{U}_F$, not $-\mathbf{U}_F$. It has been shown that using $\Phi_0$ leads to better convergence of the algorithm. However, the target propagator can be defined as $\mathbf{U}_F$ or $-\mathbf{U}_F$ leading to slightly different results.

A BURBOP-90 pulse has to be approx. $400\,\mu s$ long to excite a bandwidth of $40\,kHz$ with a quality factor of $0.99$, if the rf-amplitude is limited to $10\,kHz$. BURBOP-180 pulses have a similar relation between bandwidth and pulse duration as BURBOP-90. Possible applications for BURBOP-90 pulses include the mixing step in COSY [27], ADEQUATE [68], and INADEQUATE-type sequences [69–71], or the sensitivity-enhancement in HSQC-type experiments [72–74]. BURBOP-180 pulses have been used for refocusing of chemical shift evolution in INEPT-type sequence elements [75, 76] and in broadband TOCSY sequences [77].

### 2.5.6   ICEBERG pulses

In 2008, the GRAPE algorithm was used to design excitation pulses with a linear phase slope, the so-called ICEBERG pulses [78]. The phase slope acts as an optimization parameter $R$ and is defined as:

$$R = \frac{1}{t_p} \frac{\partial \phi}{\partial \omega} \tag{62}$$

This means that BEBOP pulses [7] are ICEBERG pulses with $R = 0$. In this work, pulses with $R$ in the range $(-1,1)$ have been optimized and characterized. It was found that $R = 0$ and $R = 1$ were the most difficult cases requiring much higher rf-amplitude for a good pulse performance. Thus, pulse duration can be significantly shorter compared to BEBOP pulses, if $R$ lies between $0$ and $1$. For example, a $39\,\mu s$ pulse with $R = 0.5$ optimized for a bandwidth of $50\,kHz$, with a tolerance to rf-miscalibration of $\pm 7\%$, achieved a quality factor of $0.99$ (rf-amplitude was limited to $15\,kHz$). A BEBOP pulse with the same bandwidth and quality factor, but without robustness with respect to rf-inhomogeneity, would have to be approximately $110\,\mu s$ long, when scaled to the same maximum rf-ampltiude.

A disadvantage compared to BEBOP pulses is that, in general, ICEBERG pulses require a 1. order phase correction. However this can be easily implemented and is a common procedure in NMR spectroscopy. The advantages are that the new pulses can cover large bandwidths up to $\Delta\nu = 10\nu_{\mathrm{rf,max}}$ and there is a significant J-coupling evolution during pulses with high $R$, which means that even longer pulses are suitable for coherence transfer applications.

# 3 Materials and Methods

## 3.1 Optimization of 30° and 60° excitation pulses

All optimizations were carried out with the software OCTOPUS [79] written in Fortran programming language. This software uses the GRAPE algorithm in conjunction with a conjugate gradient optimization method. The uncoupled-spins-approximation from section 2.4.3 is used to propagate the inital state and to back-propagate the desired target state. The states, however, are represented by quaternions.

The algorithm will stop when one of the following conditions is fullfilled:

1. The difference between old and new sum of local quality factors changes by a predefined tolerance, $\delta$.

2. The maximum number of iterations is exceeded (usually set to $2 \times 10^4$).

Different constraints were imposed during the optimization. For better comparison, rf-amplitude-limited pulses were optimized with $\nu_{\rm rf,max} = 10\,{\rm kHz}$ and in tha case of rf-power-restricted pulses, the root mean sqaure average rf-amplitude was limited to the same value.

In order to explore the physical limits of 30° and 60° pulses, the pulse performance was studied systematically in the following way: Sets of pulses were optimized for bandwidths of 10, 20, 30, 40 and 50 kHz assuming ideal rf-amplitude. In addition, sets of pulses for bandwidths of 10 and 20 kHz were calculated with rf-amplitude variations ($\vartheta = \pm 10\,\%, \pm 20\,\%$ and $\pm 40\,\%$). Pulse durations $t_p$ were varied in ranges as listed in Tables 1, 2. For each duration, 100 pulse shapes were generated randomly using a digitalization step of $0.5\,{\rm \mu s}$ and optimized using the GRAPE algorithm. The pulse with the best global quality factor for a given $t_{\rm p}$ was saved. Table 2 gives an overview of the different constraints used during the optimizations.

The rf-amplitude-variations were introduced as following. The calculation for each offset was carried out with 3 different $B_1$ values. These values were equally distributed over $\vartheta$, including 100 %. For example, for $\vartheta = \pm 20\,\%$, the $B_1$ values taken into account were 80 %, 100 % and 120 % $B_1$.

Table 1: Basic constraints used for optimizations of point-to-point pulses

| $\Delta\nu$ [kHz] | $n_{off}$ | $t_p$ [$\mu s$] | Rotation [°] |
|---|---|---|---|
| 10 | 100 | 1-200 | 30 |
| 20 | 200 | 1-200 | 30 |
| 30 | 300 | 1-200 | 30 |
| 40 | 400 | 1-200 | 30 |
| 50 | 500 | 1-200 | 30 |
| 60 | 600 | 1-200 | 30 |
| 70 | 700 | 1-200 | 30 |
| 80 | 800 | 1-200 | 30 |
| 90 | 900 | 1-200 | 30 |
| 100 | 1000 | 1-200 | 60 |
| 10 | 100 | 1-200 | 60 |
| 20 | 200 | 1-200 | 60 |
| 30 | 300 | 1-200 | 60 |
| 40 | 400 | 1-200 | 60 |
| 50 | 500 | 1-200 | 60 |
| 60 | 600 | 1-200 | 60 |
| 70 | 700 | 1-200 | 60 |
| 80 | 800 | 1-200 | 60 |
| 90 | 900 | 1-200 | 60 |
| 100 | 1000 | 1-200 | 60 |

## 3.2 Optimization of 30° and 60° universal rotation pulses

Like in the case of point-to-point pulses, sets of pulses with the same bandwidth and maximum rf-amplitude variation were optimized using different constraints. For each duration, 100 pulse shapes were generated randomly using a digitalization step of 0.5 μs and optimized using the GRAPE algorithm (see 2.4.4). The pulse with the best global quality factor for a given $t_p$ was saved. Table 2 gives an overview of the different constraints used during the optimizations.

Table 2: Basic constraints used for optimizations of universal rotation pulses

| $\Delta\nu$ [kHz] | $n_{off}$ | $t_p$ [$\mu s$] | Rotation [°] |
|---|---|---|---|
| 10 | 100 | 10-1000 | 30 |
| 20 | 200 | 10-1000 | 30 |
| 30 | 300 | 10-1000 | 30 |
| 40 | 400 | 10-1000 | 30 |
| 50 | 500 | 10-1000 | 30 |
| 10 | 100 | 10-1000 | 60 |
| 20 | 200 | 10-1000 | 60 |
| 30 | 300 | 10-1000 | 60 |
| 40 | 400 | 10-1000 | 60 |
| 50 | 500 | 10-1000 | 60 |

## 3.3 Optimization of broadband 90° universal rotation pulses with different starting shapes

When designing pulses for $^{19}$F NMR experiments, the bandwidth $\Delta\nu$ was set to $120\,\text{kHz}$. Three different approaches were used to optimize 90° UR pulses:

1. Starting with a 90° CHIRP pulse [53] with $\nu_{\text{rf,max}} = \sqrt{\frac{\Delta\nu}{t_p}}$. A sine-bell smoothing was applied at the edges of the pulse.

2. Starting with a 90° BIR-4 pulse [65, 66] with $\nu_{\text{rf,max}} = 20\,\text{kHz}$, $\Delta\omega = 2\pi\Delta\nu$, $\zeta = 10$ and $\kappa = \arctan(20)$. The number of digitalization points was always a multiple of 4 so that amplitudes and phases could be calculated according to section 2.5.2.

3. 100 random starting pulses

The pulse durations were in the range $500\text{-}2000\,\mu\text{s}$.

## 3.4 Simulations

Simulations of a single pulse on a single spin were carried out with Matlab2018b. The simulations were based on the density matrix formalism within a Hilbert space. When simulating the effects of CPMG and PROJECT sequences on two-spin systems, a Python script based on the same formalism was used.

Different excitation, universal rotation and inversion pulses optimized for a bandwidth of $120\,\text{kHz}$ have been combined to CPMG and PROJECT sequences with $\Delta = 2.5\,\text{ms}$ and comparable overall duration. PROJECT sequences were simulated with $n = 40$ and CPMG with $n = 80$. The maximum rf-amplitude was set to $20\,\text{kHz}$ for BURBOP-90 pulses and to $17.5\,\text{kHz}$ for ICEBERG, BURBOP-180 and BIBOP pulses. The pulse shapes can be found in Appendix A.

First, simulations were carried out in the absence of fluorine-fluorine coupling, then with a weak coupling assuming $J_{\text{FF}} = 15\,\text{Hz}$. The quality factor of each sequence was defined as the y-magnetization in the end of the sequence, averaged over 201 linearly distributed frequency offsets within a bandwidth of $120\,\text{kHz}$.

# 4 Results and Discussion

## 4.1 Physical limits of broadband 30° and 60° excitation pulses

### 4.1.1 Nomenclature

This chapter will show and discuss the results of various optimizations of 30° and 60° point-to-point pulses. Similar to previously published broadband excitation pulses [6, 7] that were optimized with limited rf-amplitude will be referred to as BEBOP pulses, while pulses that were optimized with limited rf-power will be referred to as power-BEBOP [8]. In addition, the net rotation angle $\alpha$ as well as pulse duration $t_p$, maximum rf-amplitude $\nu_{rf,max}$, maximum rf-amplitude deviation $\vartheta$ and bandwidth $\Delta\nu$ have to be given in order to fully describe a pulse, i. e., (power) - BEBOP - $\alpha(\Delta\nu, \nu_{rf,max}, t_p, \vartheta)$. This nomenclature is based on [9], but previously there was no need to give the angle $\alpha$ as it was always 90°.

These pulses were optimized to transfer initial $M_z$ into the transverse plane so that the final magnetization is:

$$\vec{M} = \begin{pmatrix} 0 \\ -\sin(\alpha) \\ \cos(\alpha) \end{pmatrix} \tag{63}$$

### 4.1.2 Global quality factor as a function of pulse duration

The parameters for the optimizations are listed in Table 1. In Fig. 7, the maximum global quality factors $\Phi$ of BEBOP-30 and power-BEBOP-30 pulses are plotted as a function of pulse duration.

The global quality factors of power-BEBOP-30 pulses that were optimized for a specified bandwidth $\Delta\nu$ and no rf-amplitude variation show an exponential time-dependence for short pulse durations. Similar plots were reported for 90° power-BEBOP pulses [8]. For smaller flip-angles, the quality factors for a given pulse length and bandwidth are much higher than for 90° pulses.

An additional observation is that the exponential growth is only a small part of the plots (see Fig. 7.B). The other part of the plots probably wasn't observed in the previous work because only pulses with $\Phi \lesssim 0.995$ were optimized. In contrast, this work allowed convergence to

29

higher quality factors, as long as $t_p \leq 200\,\mu s$. The second part of the plots shows the following behaviour: Once a quality factor of approx. 0.9999 is reached, there is almost no improvement in performance, but there are strong fluctuations (see Fig. 7.B). In the case of point-to-point pulses, it should always be possible to find a longer pulse with at least equal performance because the additional time could be used as a delay before the shorter pulse. Therefore, if some pulses have lower quality factors than shorter pulses with the same bandwidth, this

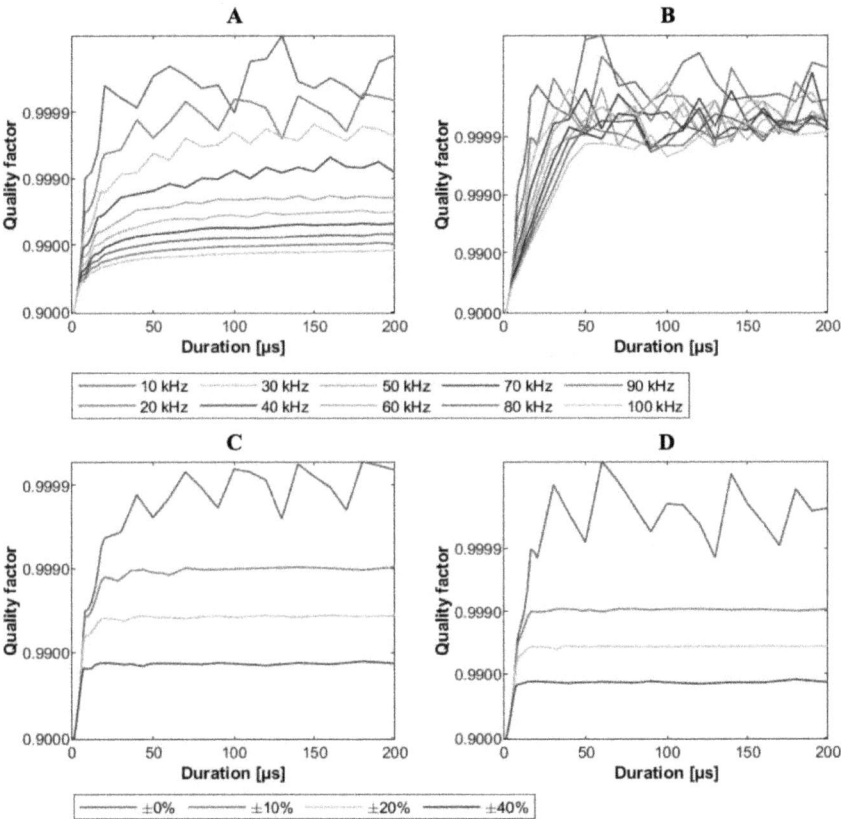

**Figure 7:** Maximum quality factors reached for BEBOP-30 (A,C) and power-BEBOP-30 (B,D) under various optimization constraints. In subfigures A and C, rf-amplitude of each pulse point is limited to 10 kHz, while in B and D, the root mean square average rf-amplitude is limited to 10 kHz. The maximum quality factor with respect to pulse duration is depicted on a logarithmic scale for ten different bandwidths $\Delta\nu$ (A,B). Subfigures C and D show maximum quality factors for a fixed bandwidth of 20 kHz and different rf-variation ranges $\vartheta$ on a logarithmic scale.

means that the algorithm did not always converge to a global minimum. This can happen when using first order or second order approximations of gradients in numerical optimizations. In order to obtain the best possible quality factors in such situations, it is neccesary to perform more optimizations with different starting shapes.

The global quality factors of BEBOP-30 pulses do not increase exponentially with pulse duration (see Fig. 7.A); the plots show a step-like behaviour for short pulse durations, similar to BEBOP-90 pulses [7]. Due to the smaller excitation angle, the plots are much steeper than for BEBOP-90 pulses. For longer pulse durations, a plateau is reached and there seem to be different limits for different bandwidths $\Delta\nu$. Compared to power-BEBOP-30, the quality factors are much lower, especially for large bandwidths. The plots achieved for BEBOP-60 and power-BEBOP-60 pulses are similar to plots of the corresponding 30° and 90° pulses (see Fig. 8). The plots are flatter than the corresponding plots of 30° pulses, but steeper than the corresponding plots of 90° pulses. These differences can be examined best, when looking at minimum pulse durations that are necessary for achieving a particular quality factor for a specific bandwidth (see Section 4.1.3).

For BEBOP-30 pulses that are robust towards rf-amplitude variation, there seem to be much

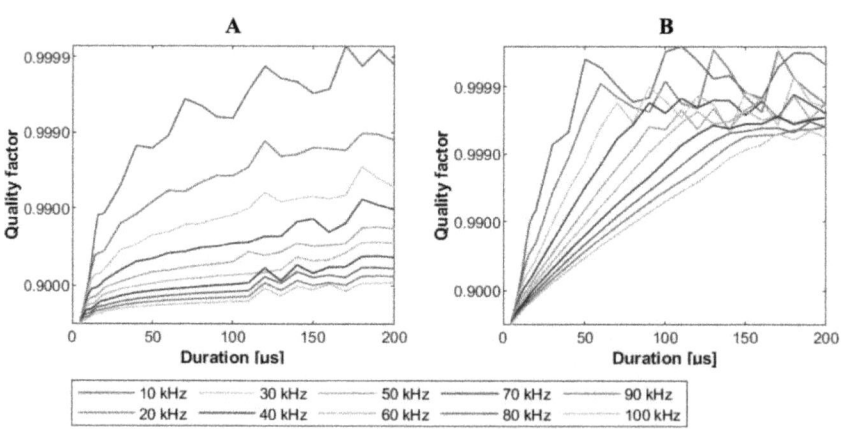

**Figure 8:** Maximum quality factors reached for BEBOP-60 (A,C) and power-BEBOP-60 (B,D) under various optimization constraints. In subfigures A and C, rf-amplitude of each pulse point is limited to 10 kHz, while in B and D, the root mean square average rf-amplitude is limited to 10 kHz. The maximum quality factor with respect to pulse duration is depicted on a logarithmic scale for ten different bandwidths $\Delta\nu$ (A,B). Subfigures C and D show maximum quality factors for a fixed bandwidth of 20 kHz and different rf-variation ranges $\vartheta$ on a logarithmic scale.

stricter physical limitations. After a particular $\Phi$ value is reached, there is no increase in performance for longer pulse durations (see Fig. 7.C and D). For a bandwidth of 20 kHz and rf-amplitude variation of $\pm 20$ %, $\Phi_{max} \approx 0.996$. Note that, if $\vartheta > 0$, power-BEBOP-30 pulses do not perform better than BEBOP-30 pulses. This is different to what was observed for 90° pulses [7,8], where the power-limited pulses perform significantly better even in these cases. Further investigations should include 60° pulses that are robust with regards to rf-amplitude variation, as it would allow a better overall comparison between BEBOP pulses of different angles.

### 4.1.3   Minimum pulse duration

In practice it is important to know which pulse duration is necessary for attaining a given quality factor over a specified bandwidth. Therefore, based on all the optimizations, the minimum pulse durations $t_{p,min}$ for a given bandwidth and quality factor were plotted for BEBOP-30, BEBOP-60, power-BEBOP-30 and power-BEBOP-60 pulses (see Fig. 9).

While for BEBOP pulses $t_{p,min}$ grows approximately exponentially with respect to bandwidth, for power-BEBOP pulses the dependence is closer to a linear relation. power-BEBOP plots have steps, which are caused by a lack of acquired data points. For example, between 10 and 20 μs the pulse duration was increased in steps of 2 μs, *i. e.*, no 11 or 13 μs pulses were optimized (see Fig. 9.B and D).

In Fig. 9.C, there are no data points for bandwidths greater than 50 kHz. This means that all the BEBOP-60 pulses optimized for these bandwidths have global quality factors $< 0.980$. In order to achieve higher quality factors, durations longer than 200 μs should be considered. Estimation of pulse performance and $t_{p,min}$ values could be made by extrapolating the plots from Figs. 8 and 9, respectively.

Note that power-BEBOP pulses can be significantly shorter in order to achieve the same performance as BEBOP pulses, especially for large bandwidths. For example, if the bandwidth is 60 kHz and the desired quality factor is 0.995, then $t_{p,min}$(BEBOP-30) $= 50$ μs and $t_{p,min}$(power-BEBOP-30) $= 14$ μs. In practice, however, maximum rf-amplitude might be limited, which means that power-BEBOP pulses may need to be scaled down to a lower $\nu_{rf,max}$ leading to a lower $\Phi$.

Using the figures from [7] and [8], a systematic $t_{p,min}$ comparison between excitation pulses with different flip-angles $\alpha$ and inversion pulses could be performed for $\Phi = 0.98$ (see Fig.10).

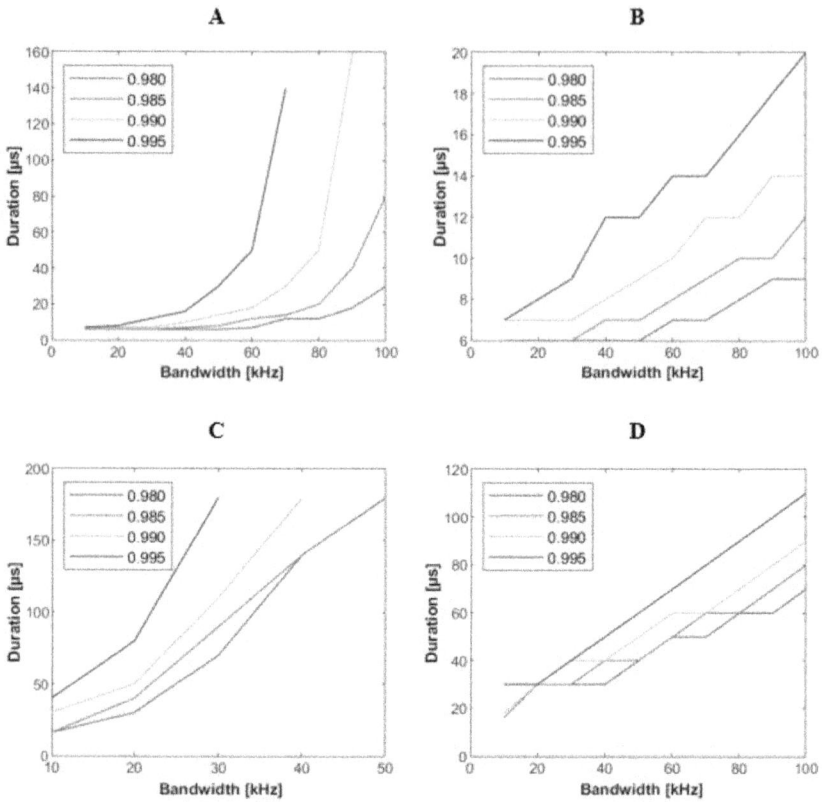

**Figure 9:** Minimum pulse length for a given bandwidth and quality factor for BEBOP-30 (A), power-BEBOP-30 (B), BEBOP-60 (C) and power-BEBOP-60 (D) pulses.

For rf-amplitude limited pulses, the slope is almost equivalent for BEBOP-90 and BIBOP pulses. For BEBOP-30 pulses, $t_{p,min}$ is almost constant and for BEBOP-60, the slope of the plots increases with bandwidth until $\Delta\nu = 40\,\text{kHz}$. In the case of power-BIBOP and power-BEBOP-30 pulses, there is almost no difference compared to the rf-power limited pulses. power-BEBOP-60 and power-BEBOP-90 pulses show a very different behaviour compared with the corresponding BEBOP pulses: The slope of the curves is much shallower which means that $t_{p,min}$ is significantly shorter.

This could be explained by comparing the transverse magnetization $M_{xy}$ of the desired target states: $M_{xy}(180°) < M_{xy}(30°) < M_{xy}(60°) < M_{xy}(90°)$. With growing $M_{xy}$, the influence of the frequency offset of a spin on its trajectory increases. As a result, it is more difficult to

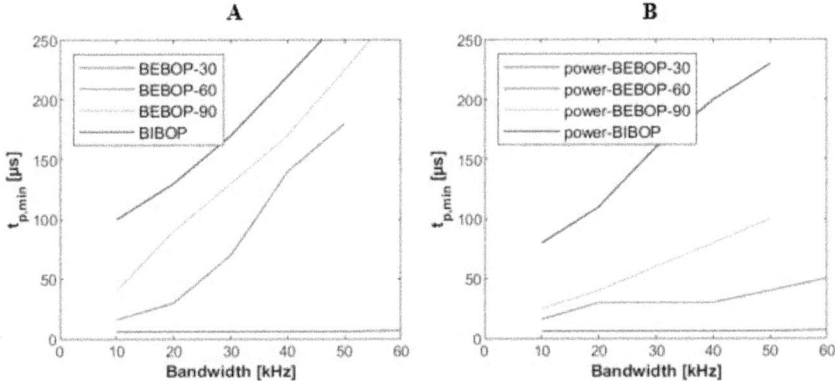

**Figure 10:** Minimum pulse length for a given bandwidth and a quality factor of 0.98 for BEBOP pulses with different flip-angles and BIBOP pulses (A) and for power-BEBOP/power-BIBOP pulses (B).

find a pulse that would lead all spins within a large bandwidth to the same target state, if the maximum rf-amplitude is limited to a value that is significantly smaller than the bandwidth. The performance of rf-power limited excitation pulses is significantly better because the peak rf-amplitude is comparable to the bandwidth $\Delta\nu$. Therefore the minimum pulse durations are shorter.

The ratios of $t_{\text{p,min}}$ for different flip-angles depend on the bandwidth and will be denoted as

$$\chi(\alpha_1, \alpha_2, \Delta\nu) = t_{\text{p,min}}(\alpha_1, \Delta\nu)/t_{\text{p,min}}(\alpha_2, \Delta\nu) \tag{64}$$

with $\alpha_1 < \alpha_2$. For a hard pulse, $\chi_{hard}(\alpha_1, \alpha_2) = \alpha_1/\alpha_2$, for example, $\chi_{hard}(30, 60) = 0.5$, $\chi_{hard}(30, 90) = 0.\overline{3}$. For the optimized pulses, $\chi < \chi_{hard}$ in most of the cases. In some cases, the values are very different from $\chi_{hard}$, especially for large bandwidths and $\alpha_1 = 30$. To give an example, for rf-amplitude-limited pulses, $\chi(30, 60)$ is in the range of $0.03 - 0.38$, but $\chi_{hard} = 0.5$.

The optimized pulses should also be compared to the best published pulses with same flip angles. But a direct comparison with the global quality factors of RADFA pulses [38,39] would be unfair because those quality factors are defined as the average performance over a range of flip-angles between 0° and 180°. However, some of the published figures showed local quality factors for different angles. Based on those figures and other information from that publication, following conclusion can be made: BEBOP pulses usually can be shorter for the same performance as RADFA pulses, but they have to be optimized for a specific angle.

On the other hand, RADFA pulses are scalable with respect to flip angles, like hard pulses. Therefore, in experiments with a known optimal flip-angle, BEBOP pulses are the best performing pulses. However, RADFA pulses are very useful, if the optimal flip-angle has to be determined. Future work will include simulations comparing the performance of different BEBOP and RADFA pulses.

Optimizing point-to-point pulses is not the only way to design excitation pulses. Pulses with a linear phase slope, like ICEBERG [78] or linear slope RADFA [39], can be used in most cases. The phase can be corrected by adjusting the delay after the pulse or applying a linear phase correction for the processing of the spectra. This additional variability during the optimization usually resulted in shorter pulses with the same performance as the corresponding point-to-point pulses. This approach should be applied in future to design ICEBERG pulses with different flip angles.

### 4.1.4 Frequency offset profiles and local quality factors

In order to assess the performance of the pulses over a desired bandwidth $\Delta \nu$, many simulations were carried out using the density matrix formalism. 501 linearly distributed offset points over an offset range of $100\,\mathrm{kHz}$ were used for calculations. The offset dependence of the final state after applying selected pulses can be seen in Fig. 11.

While the offset profiles of $M_\mathrm{x}$ are often anti-symmetric in the desired offset range, the profiles of $M_\mathrm{y}$ and $M_\mathrm{z}$ are symmetric. The local quality factors $\Phi_\mathrm{loc}$ are defined as

$$\Phi_\mathrm{loc}(\nu) = M_\mathrm{x}(t_\mathrm{p})\lambda_\mathrm{x} + M_\mathrm{y}(t_\mathrm{p})\lambda_\mathrm{y} + M_\mathrm{z}(t_\mathrm{p})\lambda_\mathrm{z} \tag{65}$$

with frequency offset of the spin $\nu$ and desired target state $\lambda$.

Fig. 12 shows a comparison of local quality factors of selected BEBOP-30 and power-BEBOP-30 pulses with bandwidths $\Delta \nu = 10\,\mathrm{kHz}$ and $50\,\mathrm{kHz}$, respectively. A major observation is that not just the global quality factor, but also the pulse selectivity increases with duration. Rf-power-limited pulses with same $\Delta \nu$ and $t_\mathrm{p}$ are less selective than rf-amplitude-limited pulses, especially for small pulse durations and bandwidths.

Very short BEBOP-30 and power-BEBOP-30 pulses (e. g. $t_\mathrm{p} = 5\,\mathrm{\mu s}$) have the same offset

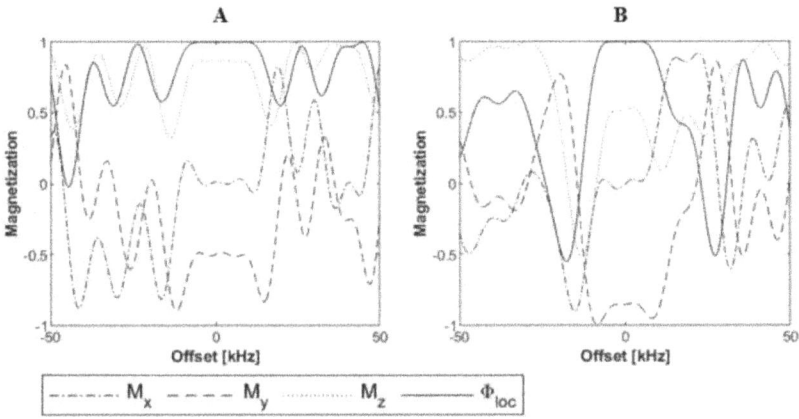

**Figure 11:** Frequency offset profiles of $M_x$, $M_y$, $M_z$ and $\Phi_{loc}$ after a $100\,\mu s$ BEBOP-30 (A) /BEBOP-60 pulse (B). Both pulses were optimized for a bandwidth of $10\,kHz$ and without rf-amplitude variation.

profile for all bandwidths. An analysis of pulse shapes showed that all these pulses have the same shape: they are hard pulses. A detailed report on pulse shapes is subject of the next subsection.

The local quality factors of BEBOP-60 and power-BEBOP-60 pulses depend stronger on the pulse duration (see Fig. 13). This fits well with the relatively moderate increase of global $\Phi$ with $t_p$ reported in the previous section. Shorter pulses have very symmetric offset profiles, while the profiles of longer pulses are not symmetric and have strong variations outside of the desired offset range. Again, the selectivity grows with pulse duration and is higher for rf-amplitude-restricted pulses.

The longer the pulse, the more small oscillations are visible within the desired bandwidth but the deviations of the local quality factors from the global quality factor become smaller (see Fig. 13.B). These deviations are not noticable for small bandwidths, but they are on the scale of $10^{-2}$ for a bandwidth of $50\,kHz$. As a consequence, longer pulse durations should be used for a more uniform behaviour over large bandwidths. In further work, pulse durations up to $500\,\mu s$ for BEBOP-60 pulses should be considered.

### 4.1.5 Peak rf-amplitudes and pulse shapes

The superior performance of power-BEBOP pulses is probably achieved through a higher maximum rf-amplitude. The relation between peak rf-amplitude $\nu_{\text{rf,max}}$ and pulse duration $t_p$ is shown in Fig. 14. Power-BEBOP-30 pulses show a slightly different behaviour to power-BEBOP-60 and to previously published power-BEBOP-90 pulses [8]. For short pulse durations, $\nu_{\text{rf,max}}$ increases approximately linearly with respect to pulse duration until it reaches a peak. Then it goes to a slightly lower value and fluctuates for long pulse durations.

For power-BEBOP-60 pulses, the relation is more similar to the previously published power-BEBOP-90 pulses. For most of the observed pulses $\nu_{\text{rf,max}}$ grows linearly with increasing pulse

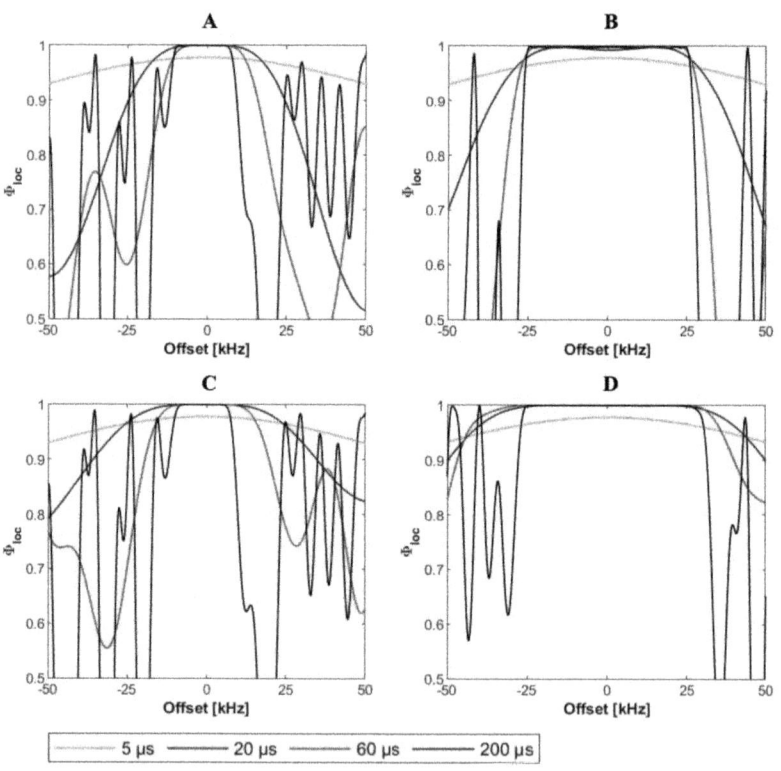

**Figure 12:** The local quality factors are depicted for BEBOP-30 (A,B) and power-BEBOP-30 pulses (C,D). In all subfigures same colors correspond to the same pulse durations. In A and C the desired bandwidth was 10 kHz, in B and D - 50 kHz.

duration, and for large bandwidths, the maximum $\nu_{\text{rf,max}} \approx \Delta\nu$. Note that in the study on power-BEBOP-90 pulses, a smaller number of pulses with high quality factors have been examined because of the optimization set up: For a given bandwidth, the pulse length was increased until $\Phi > 0.995$. Because of this, it is possible that power-BEBOP-90 pulses also show the same behaviour, but only the linear parts of the plots have been observed.

The pulse shapes revealed that there are three families of BEBOP-60 pulses: For short durations, hard pulses were obtained from optimizations. The second family are phase-alternating pulses. For $\Delta\nu = 50\,\text{kHz}$, these have pulse durations in the range of 12 - 30 µs. For example, one pulse roughly corresponds to $20°_{-x}$, $52°_x$ (see fig. 15B,C). For a bandwidth of 10 kHz, the

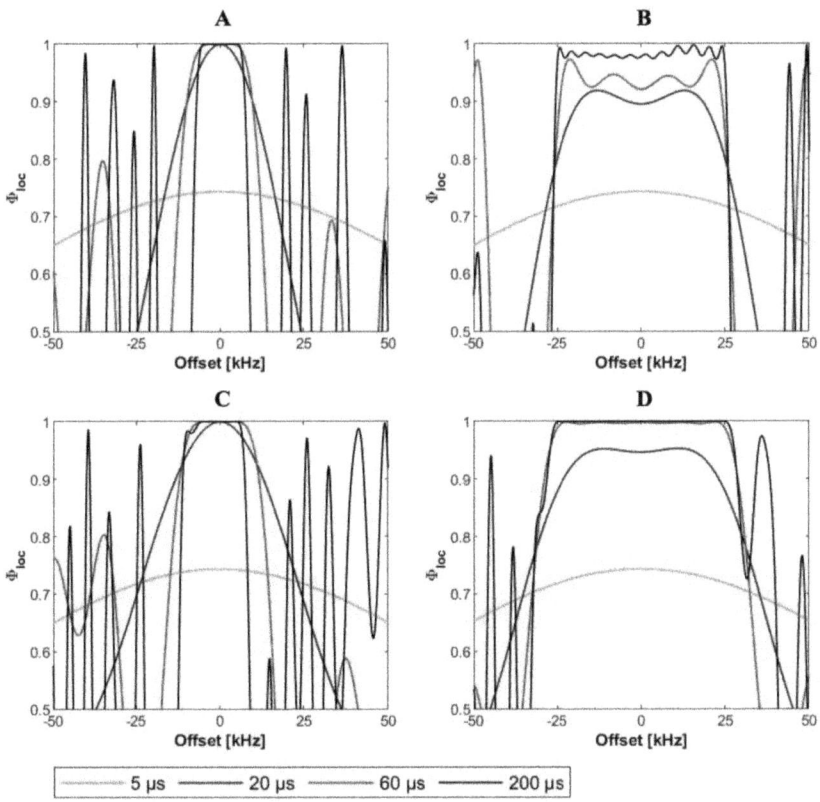

**Figure 13:** The local quality factors are depicted for BEBOP-60 (A,B) and power-BEBOP-60 pulses (C,D). In all subfigures, same colors correspond to the same pulse durations. In A and C the desired bandwidth $\Delta\nu = 10\,\text{kHz}$; in B and D, $\Delta\nu = 50\,\text{kHz}$.

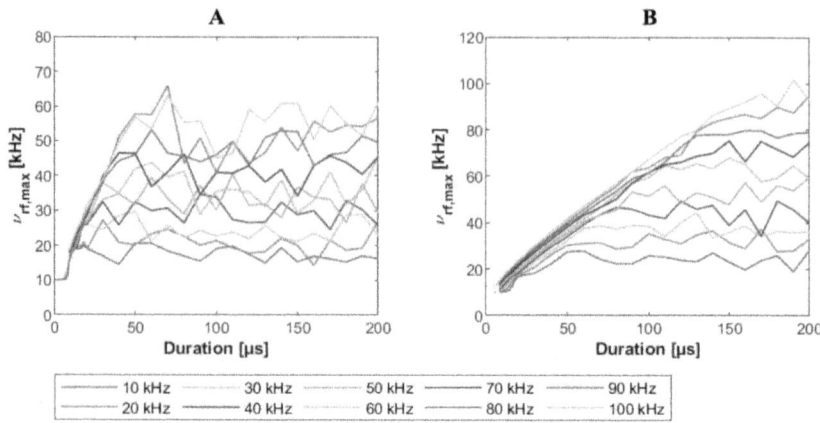

**Figure 14:** Peak rf-amplitude of power-BEBOP-30 (A) and power-BEBOP-60 pulses (B) as a function of pulse duration for different bandwidths.

phase-modulations start at a duration of $18\,\mu s$. Pulses that belong to the third pulse family have unique shapes because y-amplitude is modulated simultaneously with x-amplitude. For very long pulse durations, the shapes are very noisy. This is evidence of incomplete optimization. There are significant amplitude modulations in the first half of the pulses, while the maximum allowed rf-amplitude is used for the whole second half of the pulses (see fig. 15D).

All power-BEBOP-30 pulses seem to belong to the same pulse family, with a sinc-like amplitude modulation and simultaneous phase jumps. These shapes are very similar to power-BEBOP-90 pulses [8], polychromatic pulses [80], E_BURP pulses [81] and ICEBERG pulses [78].

The number of phase-shifts depends on the pulse duration $t_p$ as well as the bandwidth $\Delta\nu$ and rf-amplitude variation $\vartheta$. For a bandwidth of $10\,kHz$ and no rf-variation, pulses shorter than $20\,\mu s$ have no phase-modulation. However, for a bandwidth of $50\,kHz$, the first phase jump is introduced when the pulse length equals $12\,\mu s$. The maximum amplitude is always reached in the end of the pulse.

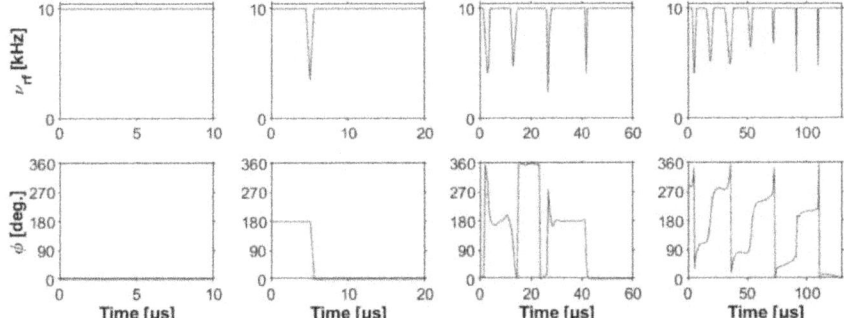

**Figure 15:** Amplitude and phase of selected optimized 60° excitation pulses with limited amplitude found for a bandwidth of 50 kHz and no rf amplitude variation. The pulses in A and B roughly correspond to $36°_x$ and $20°_{-x}$, $52°_x$

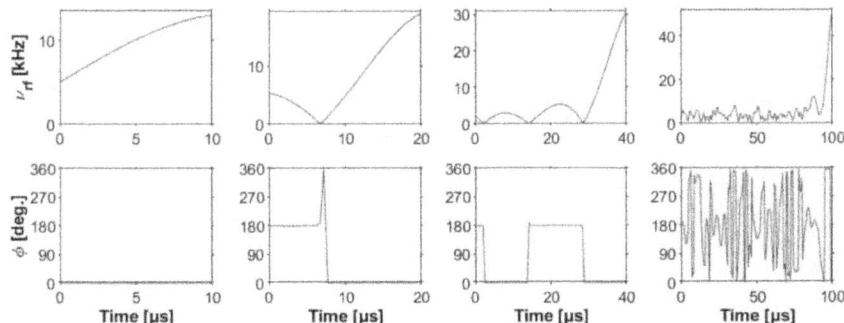

**Figure 16:** Amplitude and phase of selected optimized 60° excitation pulses with limited power deposition found for a bandwidth of 50 kHz and no rf amplitude variation. The maximum rf amplitude rises with pulse duration and is always reached in the end of the pulse.

40

## 4.2 Broadband 30° and 60° universal rotation pulses

### 4.2.1 Nomenclature

This section will show and discuss the results of optimizations of 30° and 60° universal rotation pulses. Pulses optimized with limited rf-amplitude can be described using the published nomenclature for BURBOP pulses [9]. A similar nomenclature can be used for pulses that were optimized with limited rf-power, which will be referred to as power-BURBOP pulses. The pulses can be fully described by the following parameters: rotation angle $\alpha$, rotation axis $\xi$, maximum rf-amplitude $\nu_{max}$, the pulse length $t_p$, bandwidth $\Delta\nu$ and maximum rf-amplitude deviation $\vartheta$, i. e., (power) - BURBOP - $\alpha_\xi(\Delta\nu, \nu_{\text{rf,max}}, t_p, \vartheta)$.

### 4.2.2 Global quality factor and pulse duration

The convergence of UR pulse optimizations is much slower than of PP pulses. Therefore, considering available computation resources, only bandwidths up to $50\,\text{kHz}$ have been used in this study (see table 2, section 2). It has been shown previously by Kobzar et al. [9] that the algorithm converges better if there is no ambiguity in the quality factor $\Phi$. This means that the same phase factor $e^{i\phi}$ has to be used for all offsets. For a spin $1/2$, this factor is either +1 or -1. First, $e^{i\phi} = -1$ was used for all optimizations.

The best global quality factors for BURBOP-30 and power-BURBOP-30 pulses reached with this approach are depicted on a logarithmic scale in Fig. 17. For very short pulse durations, the best global quality factors are negative. A reason for this could be that a negative phase factor ($e^{i\phi} = -1$) was used to optimize $U_F$, which corresponds to a $330_{-x}$ pulse instead of a $30_x$ pulse and a $300_{-x}$ pulse instead of a $60_x$ pulse.

In addition, the same optimizations were performed using a positive phase factor ($e^{i\phi} = 1$). This approach always provided significantly better solutions for very short pulses, Fig. 18 shows a comparison of best $\Phi$ achieved with different phase factors for a bandwidth of $50\,\text{kHz}$. The best solutions can only be achieved by using both approaches, as they alternately provide best results depending on the pulse duration. For long pulses ($t_p > 400\,\mu\text{s}$), there is no noticable difference between the two plots.

Almost all the plots show an oscillating behavior that has also been observed for BURBOP-90 pulses [9]. The explanation given by Kobzar et. al. in that paper seems to be applicable to all

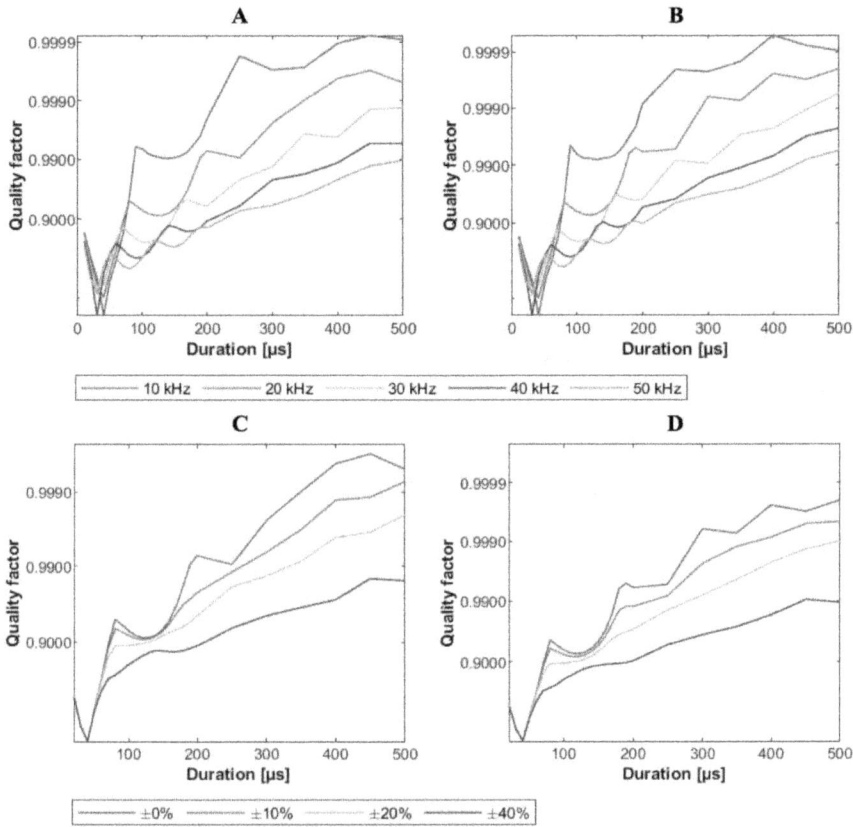

**Figure 17:** Maximum quality factors reached for BURBOP-30 (A,C) and power-BURBOP-30 (B,D) under various optimization constraints. In subfigures A and C, rf-amplitude of each pulse point is limited to 10 kHz, while in B and D, the root mean square average rf-amplitude is limited to 10 kHz. The maximum quality factor with respect to pulse duration is depicted on a logarithmic scale for five different bandwidths $\Delta\nu$ (A,B). Subfigures C and D show maximum quality factors for a fixed bandwidth of 20 kHz and different rf-variation ranges $\vartheta$ on a logarithmic scale.

UR pulses with $\alpha_\zeta \neq 180°$: A UR pulse with equal performance cannot be constructed from a shorter UR pulse by simply adding a delay in the beginning of the pulse because a delay results in larger chemical shifts that have to be compensated by the UR pulse. For this reason, in some cases, the performance will decrease with pulse duration. For $t_p < 400\,\mu s$, power-BURBOP-30 pulses perform better than BURBOP-30 pulses, but there is no significant difference for longer pulse durations.

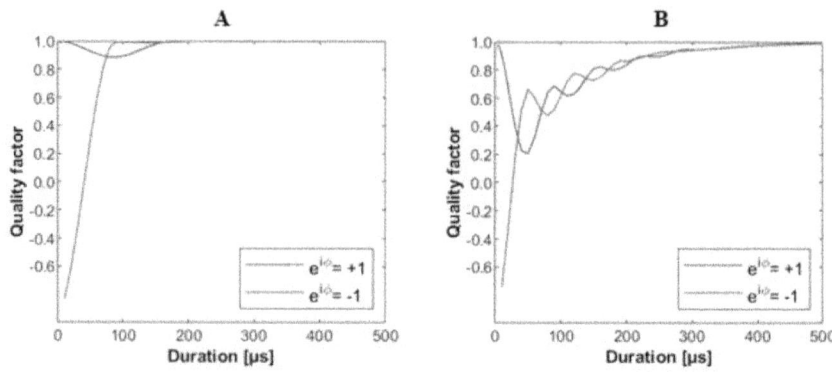

**Figure 18:** Maximum quality factors reached for BURBOP-30 pulses with $\Delta\nu = 10\,\mathrm{kHz}$ (A) and $50\,\mathrm{kHz}$ (B) using different phase factors.

The global quality factors achieved for 60° UR pulses show a similar behaviour, but there is a slower increase in $\Phi$ with respect to pulse duration. Like in the case of PP pulses, the

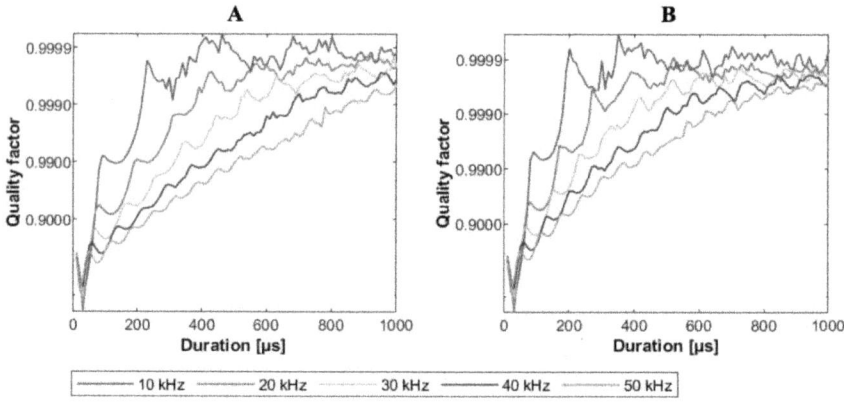

**Figure 19:** Maximum quality factors reached for BURBOP-60 (A) and power-BURBOP-60 (B) under various optimization constraints. In A, rf-amplitude of each pulse point is limited to $10\,\mathrm{kHz}$, while in B, the root mean square average rf-amplitude is limited to $10\,\mathrm{kHz}$. The maximum quality factor with respect to pulse duration is depicted on a logarithmic scale for five different bandwidths $\Delta\nu$.

minimum pulse length as a function of bandwidth and quality factor is of high practical interest. In Fig. 20, minimum $t_\mathrm{p}$ are plotted against the desired bandwidth $\Delta\nu$ for four $\Phi$ values and different pulse sets (BURBOP-30, power-BURBOP-30, BURBOP-60, power-BURBOP-60). For all pulse sets, the relation between $\Delta\nu$ and $t_\mathrm{p}$ is roughly linear for the investigated bandwidths. This has also been reported for BEBOP, BIBOP, BURBOP-90 and BURBOP-180 pulses. For $\Delta\nu = 10$ or $20\,\mathrm{kHz}$, there is only one significant difference between the four pulse sets: In order to achieve BURBOP pulses with $\Phi = 0.995$, much higher pulse durations are necessary than for power-BURBOP pulses. All the other values are very similar and in some cases, the values cannot be given with a high precision, as for $t_\mathrm{p} \geq 200\,\mathrm{\mu s}$ only multiples of 50 were used as pulse durations. A comparison of $t_\mathrm{p,min}$ for different flip-angles $\alpha$ and $\Phi = 0.99$ showed that minimum pulse durations were roughly equal. This could be different if a much higher quality factor is desired, however, $t_\mathrm{p,min}$ data for other $\Phi$ values were not provided for BURBOP-90 and BURBOP-180 pulses [9].

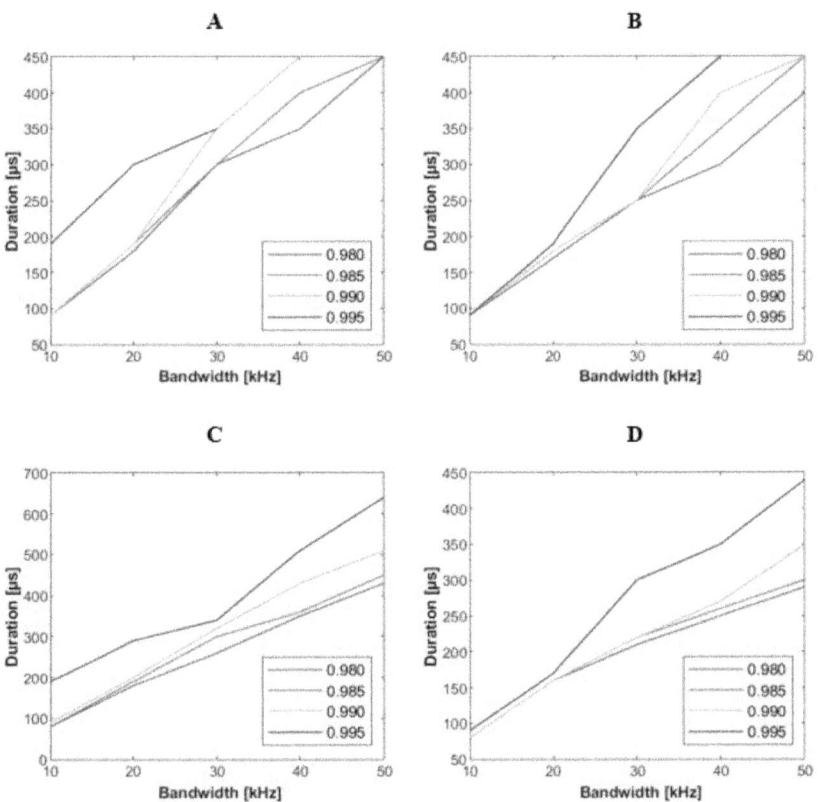

**Figure 20:** Minimum pulse durations for $\Phi = 0.980, 0.985, 0.990$ and $0.995$ are plotted as functions of the desired bandwidth for different pulse sets.(A): BURBOP-30, (B): power-BURBOP-30, (C): BURBOP-60, (D): power-BURBOP-60 pulses.

### 4.2.3  Frequency offset profiles and local quality factors

Fig. 21 shows a comparison of local quality factors of selected BURBOP-30 and power-BURBOP-30 pulses with $\Delta\nu = 10\,\mathrm{kHz}$ and $50\,\mathrm{kHz}$. The profiles of the correspoding 60° pulses are very similar, and therefore not shown here. BURBOP pulses seem to be more selective than BE-BOP pulses: The offset profiles outside of the desired bandwidth are very noisy, and some local quality factors have negative values. The performance within $\Delta\nu$ is very good for small bandwidths and relatively long pulse durations ($t_p \geq 400\,\mu s$). However, the deviations from a good performance are enormous ($\Phi_{loc} < 0.5$) when the global quality factor is below 0.995, which corresponds to BURBOP-30 pulses with durations below $450\,\mu s$ for $\Delta\nu = 50\,\mathrm{kHz}$ (see Fig. 21.B). The corresponding power-BURBOP pulses show a very similar behaviour, but the deviations are slightly smaller.

### 4.2.4  Peak rf-amplitudes and pulse shapes

The relation between peak rf-amplitude $\nu_{rf,max}$ and pulse duration $t_p$ of power-BURBOP pulses is shown in Fig. 22. The plots for power-BURBOP-30 and power-BURBOP-60 look similar: First, there is a step-like increase in $\nu_{rf,max}$ with increasing pulse length. Then there are strong fluctuations (see Fig. 22.B), these cannot be seen for the 30° pulses, because less different $t_p$ values were used.

The shapes of all UR pulses have been examined and compared. Most of the pulses have symmetric shapes, even though no symmetry constraints were used in the optimizations. Almost all of these pulses belong to symmetry type I [9], which means that both x- and y-amplitudes are symmetric with respect to the center of the pulse.

There are major differences between BURBOP and power-BURBOP pulses: For most BUR-BOP pulses, the rf-amplitude is close to the maximum allowed value for most of the time. In contrast, most power-BURBOP pulses have peak amplitude in the beginning, the center and the end of the pulse (see Fig. 23 and 24).

Three families of BURBOP-60 pulses were found. For very short pulse durations, hard pulses were found to be the best pulses, like in the case of BURBOP-90 and BURBOP-180 [9]. For longer durations, the algorithm delivered symmetric, relatively smooth phase-alternating pulses (see Fig. 23.B and C). Pulses that are longer than $190\,\mu s$ belong to the third family: their

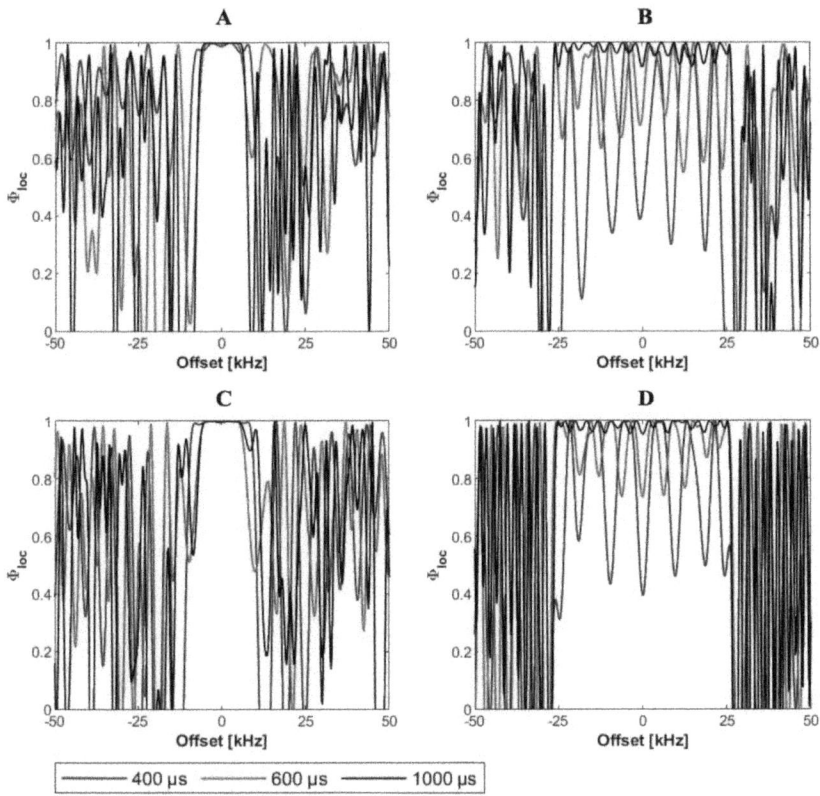

Figure 21: The local quality factors are depicted for BURBOP-30 (A,B) and power-BURBOP-30 pulses (C,D). In all subfigures, same colors correspond to the same pulse durations. In A and C the desired bandwidth $\Delta\nu = 10$ kHz; in B and D, $\Delta\nu = 50$ kHz. Only positive values of $\Phi_{loc}$ are shown.

shapes are not symmetric, but they can be divided into two almost identical sections of equal length (see Fig. 23.D). These pulses have a larger range of pulse phases ($0° < \phi < 360°$) compared to the pulses of the second family ($0° < \phi < 180°$ or $180° < \phi < 360°$).

The power-BURBOP-60 pulses could also be divided into three families: Hard pulses were found as the optimal solutions for 10 kHz and 20 kHz and short durations ($< 90$ µs). For larger bandwidths, none of the optimized pulses were hard pulses, even short pulses show amplitude modulations: The optimal solutions are symmetric, phase-modulated pulses with a sinc-like amplitude-modulation (see Fig. 24.A, B and C). These shapes are similar to RE-BURP [81] and ICEBERG pulses [78]. The longer the pulse, the more phase modulation. For very long pulse

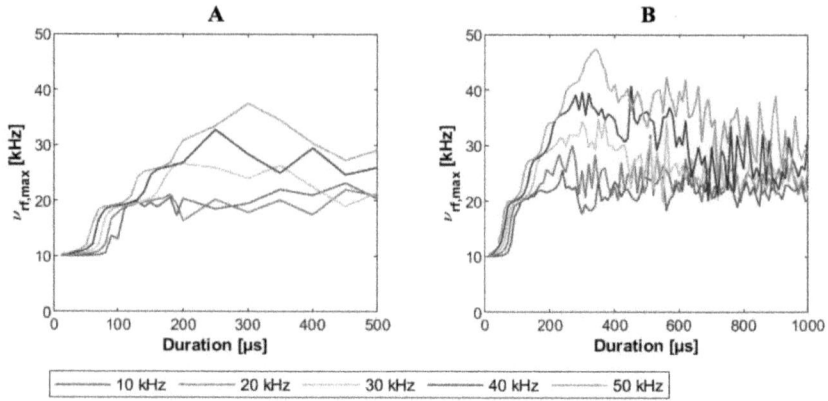

**Figure 22:** Peak rf-amplitude of power-BURBOP-30 (A) and power-BURBOP-60 (B) pulses as a function of pulse duration.

durations, the algorithm found unique-looking pulse shapes. The majority of these shapes can be divided into two almost identical sections, like the long BURBOP-60 pulses.

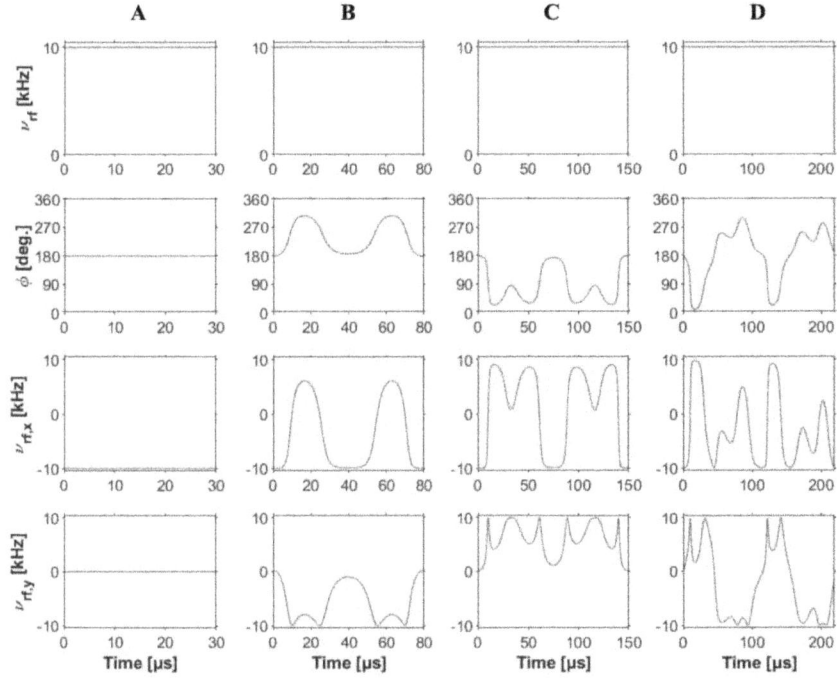

**Figure 23:** Shapes of selected BURBOP-60 pulses with a bandwidth of 50 kHz and no rf-amplitude variation: The rf-amplitude as well as the phase and the x- and y-amplitude components are plotted as functions of time.

The pulse shapes achieved for long pulse durations were very noisy. This is evidence of incomplete optimization: the global maximum wasn't found. Often, such pulses can be used for a second optimization round with a smaller $\delta$. The resulting pulses have a slightly higher quality factor and a much smoother shape.

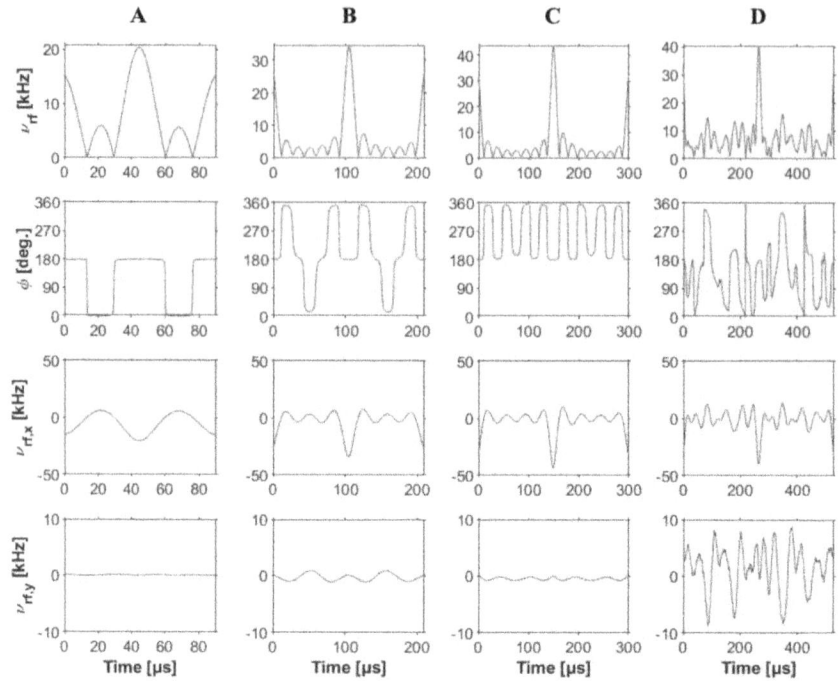

**Figure 24:** Shapes of selected power-BURBOP-60 pulses with a bandwidth of 50kHz and no rf amplitude variation. The rf-amplitude as well as the phase and the x- and y-amplitude components are plotted as functions of time. Note that the peak amplitudes are different.

## 4.3 Design of broadband CPMG and PROJECT sequences

### 4.3.1 Optimization of broadband 90° UR pulses

When scanning large libraries of fluorinated organic compounds, pulses are required to cover a bandwidth of 120 kHz. Based on previously published BURBOP pulses with smaller bandwidths [9], an estimation of the minimum pulse length could be made. Starting with $t_p = 500\,\mu s$, the pulse duration was increased stepwise to $2000\,\mu s$. The rf-amplitude was limited to 20 kHz. The GRAPE algorithm can be used with different starting pulse shapes. Here, three different approaches were used to optimize 90°UR pulses:

1. Starting with a 90° CHIRP pulse [53] (more details in Section 3.3)

2. Starting with a 90° BIR-4 pulse [65, 66] (more details in Section 3.3)

3. 100 random starting pulses

In the first two cases, two different target propagators $U_F$ corresponding to a $-270°_x$ and $-270°_y$ rotation were used. In the case of CHIRP pulses, the choice of the smoothness factor can also effect the result. The smoothness factor defines how many percent of the pulse shape points are used for smoothing, $i.\ e.$, sm=10 means 10 %. The average computation time per optimization was approximately the same for different starting pulse shapes. This means that the convergence of all approaches is comparable, but the third approach requires about 100 times longer because of the number of optimizations.

The performance of CHIRP-based pulses strongly depends on the pulse duration (see Fig. 25.A and B). The performance increases exponentially between $500\,\mu s$ and $900\,\mu s$, then the increase is slower. When $t_p = 1800\,\mu s$, there is a large jump. $\Phi$ is only slightly different, if the target propagator or the smoothness factor is changed. For pulse durations less than $1400\,\mu s$, $U_F \equiv -270°_y$ usually delivered slightly better results, for longer pulse durations the other way around. With sm=10, the algorithm found significantly better pulses for $t_p = 1000\,\mu s$; $1500\,\mu s$, but a significantly poorer performing pulse for $t_p = 2000\,\mu s$. The corresponding pulse shapes are very similar - often, the only differences are the rf-amplitudes in the beginning and the end of the pulse.

When generating CHIRP pulses as described in Section 3.3, the maximum rf-amplitude is usually lower than $20\,kHz$. During the optimization, the amplitude was increased and modulated severely in most cases. For $t_p < 750\,\mu s$, most pulses have a small area without amplitude modulations in the middle, but longer pulses only sometimes have this area (see Fig. 26.B). The phase modulations often remind of the original CHIRP shapes, but are noisier. Most pulses are symmetric around the center of the pulse. For $t_p \geq 1800\,\mu s$, the amplitude oscillates around $10\,kHz$ and the phase modulation is very similar to the modulation of the starting pulse (see Fig. 26.C).

In the case of BIR-4 based pulses, it seems to play a major role, which target propagator was used. For some pulse durations, x-pulses yield better performance; for other durations, y-pulses perform better. Very good solutions for x-rotations are achieved for pulse durations of 800, 850, 1000, 1500, 1700 and $2000\,\mu s$ (see Fig. 25.C). In contrast, pulses with a duration of 950 and $1100\,\mu s$ have particularly low global quality factors. Especially well performing BIR-4 based y-pulses have pulse durations of 550, 750, 950, 1100, 1300, 1500, 1600 and $1900\,\mu s$.

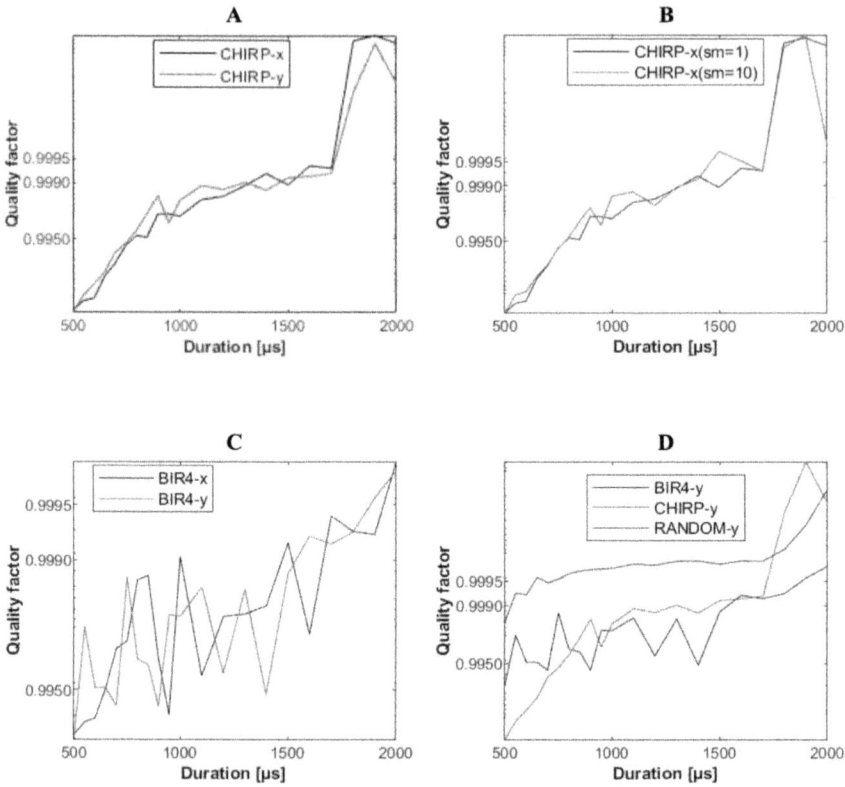

**Figure 25:** Comparison of the best global quality factors achieved with different optimization approaches using a logarithmic scale for $\Phi$. Different starting pulse shapes were used as well as different target propagators ($-270°_x$ or $-270°_y$) and different smoothness factors for CHIRP pulses. In A, CHIRP is used as the starting pulse shape with different target propagators. In B, CHIRP pulses were optimized as $90°_x$ pulses with different smoothness factors. In C, BIR-4 pulses were used as the starting pulse shapes. D shows a comparison of CHIRP-y(sm=1), BIR4-y and RANDOM-y pulses. The maximum rf-amplitude was limited to 20 kHz during the optimizations.

Pulses with durations of 700, 900 and 1400 µs have relatively low quality factors.

In many cases, very good performance seems to correspond to characteristic phase modulations making the pulse similar to the original BIR-4 shape. Poor performance can be associated with substantial time intervals with $\nu_{rf} < \nu_{rf,max}$ in most of the cases. In addition, there is an overall trend that the quality factors grow with pulse duration, like for the other pulse families. Fig. 27 shows the shapes of the 900 µs starting pulse and the resulting x-pulse and y-pulse shapes. In this case, the shape of the x-pulse is more similar to the original shape resulting in

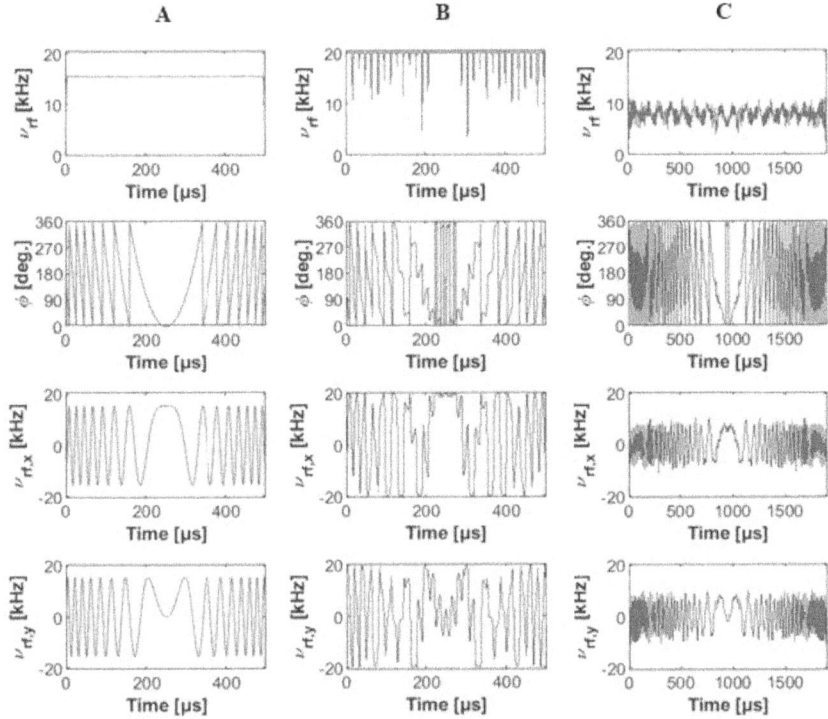

**Figure 26:** Starting shape of the 500 µs CHIRP pulse (A), the optimized shape for the same pulse duration (B) and the optimized shape of a particularly well performing pulse with $t_p = 1900\,\mu s$ (C). All these pulses were optimized with a $-270°_y$ target propagator and a smoothness factor of 1.

better performance.

A comparison of the three approaches, when optimizing a $90°_y$-rotation, is shown in Fig. 25.D. In most cases, optimizations based on random pulses yield the best results. This is interesting because the corresponding pulse shapes are seemingly noisy, showing no symmetries. The shortest pulse with $\Phi \geq 0.999$ has a pulse duration of 600 µs and will be one of the pulses used in the sequences.

Only for pulse durations of 1800 and 1900 µs, CHIRP based pulses perform better. Such long pulses should not be used for the CPMG sequence, because relaxation during the pulse - that has not been included in the optimizations - could lead to significant intensity loss in the experiment. In the PROJECT sequence, undesired coupling evolution could take place during the time of the pulse.

In this work only one CHIRP starting shape was used for each specified pulse duration. It is conceivable that using many CHIRP shape with different maximum rf-amplitudes and smoothness factors as starting pulses could lead to better results and should be subject to further investigations. Similarly, many BIR-4 pulses could be generated using different values for $\zeta$ and $\kappa$. In this approach, simply the values from [66] were used.

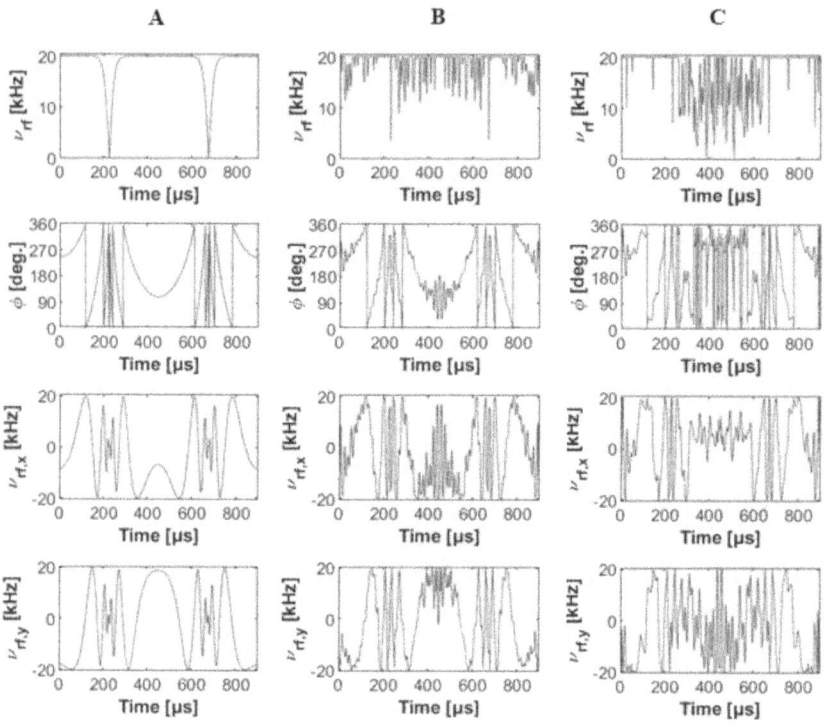

**Figure 27:** Starting shape of the 900 µs BIR-4 pulse (A) and the optimized shapes with a $-270°_x$ target propagator (B) and a $-270°_y$ target propagator (C).

### 4.3.2 Optimization of 180° PP pulses

In both sequences - CPMG and PROJECT - the refocusing pulses can be replaced by two inversion pulses (see 29). As mentioned previously, the bandwidth for $^{19}$F excitation is 120 kHz. Consequently, all pulses used in the sequence should be optimized for this bandwidth. Based on previously published BIBOP pulses [7], it was estimated that the shortest pulse covering this bandwidth with a reasonably high quality factor and allowing for $\pm 5\% B_1$ deviation would be 500 μs long. First, pulses with durations of 500 μs, 550 μs and 600 μs were optimized. The best quality factors were 0.9954, 0.9975 and 0.9979, respectively. However, for a practical application, $\Phi$ should be at least 0.999.

In order to find a better pulse, 500 further optimizations with different random seeds were started. The best $\Phi$ for the 600 μs pulse was 0.9983, which is only an improvement of 0.0004; For $t_p = 550$ μs, the improvement was only 0.0003. Consequently, further increasing the number of optimizations is not sufficient. Changing the optimization method has been considered, but implementing a new method in the software would have taken too much time when the strict time limits of this work were taken into account. When analyzing why the optimization has stopped, though the optimal solution hasn't been reached, there are two possibilities: The algorithm reached a very plane area of the cost function, where the gradients are very small, and the change of the cost function between two iterations is easily smaller than the set convergence parameter $\delta$. Another possibility is that the algorithm got trapped in a local minimum. To help with both problems, the following optimization scheme was implemented:

1. Add noise to the best pulse from the previous optimization round. This can be done using a random noise generator, resulting in many new pulses that only slightly differ from the original pulse.

2. Use these pulses as starting pulses for new optimizations with a smaller convergence parameter $\delta$.

3. Select multiple pulses with the best performance and go back to step 1.

Table 3: Results of multiple optimization rounds of BIBOP pulses

| Opt. round | N_opt | $\delta$ | $\Phi_{best}$ |
|---|---|---|---|
| 1 | 500 | $10^{-5}$ | 0.997758 |
| 2 | 50 | $10^{-8}$ | 0.998364 |
| 3 | 150 | $10^{-8}$ | 0.998390 |
| 4 | 250 | $10^{-10}$ | 0.998479 |
| 5 | 100 | $10^{-10}$ | 0.998496 |

The best quality factors after each optimization round of the 550 μs pulses are listed in Table 3. This new approach led to a pulse with a $0.000738 = 0.07$ % higher quality factor. The improvement due to multiple optimization rounds is best illustrated by comparing the inversion profiles of the 550 µs pulses (see Fig. 28). There is a clear difference between the first pulse and the second pulse. But the performance of other pulses is roughly the same as of the second pulse.

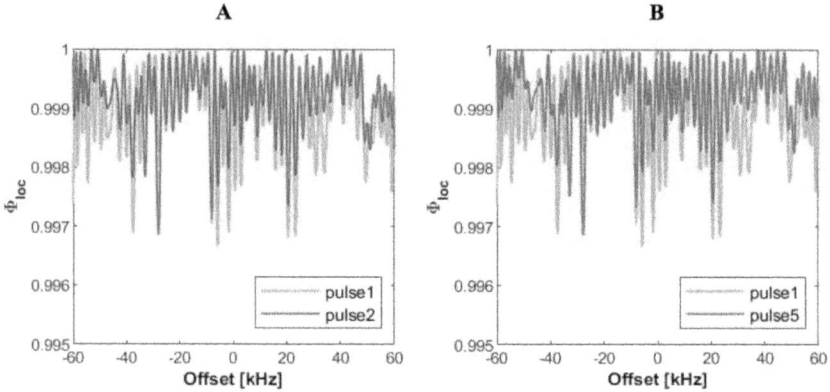

Figure 28: Comparison of the inversion profiles of the best pulses of opt. rounds 1 (green), 2 (red) and 5 (purple).

Thus, it seems that choosing $\delta \leq 10^{-8}$ does not lead to noticably better results. On the other hand, it would be good to investigate, whether inserting additional optimization rounds between round1 and round2 with $\delta = 10^{-6}$ and $\delta = 10^{-7}$ would result in a significantly better pulse. Other parameters that could be modified are the number of best pulses selected after each optimization round, the amount of noise added to them as well as the number of resulting pulses (with noise). If a better pulse exists, exact gradient based algorithms might be able to find it, but it would require additional time to implement these methods into the OCTOPUS software.

### 4.3.3  Simulation of different CPMG and PROJECT sequences

Different pulses that were optimized for this master thesis ?? as well as other pulses available were combined to form CPMG and PROJECT sequences (see Fig. 29). The pulse shapes can be found in Appendix A. These pulse sequences have been simulated under different conditions: In the absence of scalar coupling and with a weak coupling ($J_{FF} = 15$ Hz). In all cases, the delay was 2.5 ms. The number of echoes was 80 for CPMG and 40 for PROJECT to ensure that both experiments had approximately the same duration. The maximum rf-amplitude was set to 20 kHz for BURBOP-90 pulses and to 17.5 kHz for ICEBERG, BURBOP-180 and BIBOP pulses. The quality factor of the sequences was defined as the y-magnetization in the end of the sequence averaged over 201 linearly distributed frequency offsets within a bandwidth of 120 kHz.

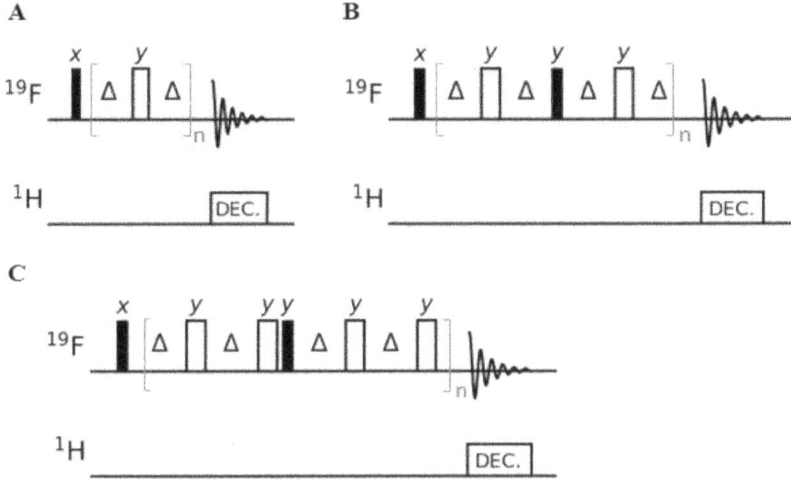

**Figure 29:** The standard CPMG sequence (A), the PROJECT sequence (B) and the modified PROJECT sequence that was used with 4 BIBOP pulses per echo that are illustrated by empty rectangles (C).

**Simulations in the absence of scalar coupling**

For the CPMG sequence, the best results were achieved when the BURBOP-90 pulse with ±10 % robustness was used for excitation (see Table 4). Depending on the refocusing pulse(s), $\Phi$ was in the range of $0.9979 - 0.9984$. The other BURBOP pulse and the ICEBERG pulse had only slightly lower quality factors. The best refocusing pulse was the quasi-adiabatic BURBOP pulse, closely followed by the symmetric and anti-symmetric BURBOP pulses. The use of 2 BIBOP pulses for refocusing resulted in significantly lower quality factors ($0.5194 - 0.5200$).

In comparison to the shaped pulses, hard pulses showed very poor performance. When a hard pulse was used for excitation over a bandwidth of 120 kHz, then the global quality factors were in the range of $0.2394 - 0.2418$ depending on the used refocusing pulse. In NMR experiments, a 1. order phase correction can be used to reduce the phase errors. This was not implemented in the simulations, but in the case of the ICEBERG pulse the delay after the pulse

Table 4: Global quality factors $\Phi$ of different CPMG pulse sequences in the absence of J coupling and relaxation

| Excitation pulse | Refocusing pulse | $\Phi$ |
|---|---|---|
| Hard 90°$_x$ | asym. | 0.2394 |
| Hard 90°$_x$ | sym. | 0.2418 |
| ICEBERG | asym. | 0.9965 |
| ICEBERG | sym. | 0.9970 |
| ICEBERG | quasi-adiab. | 0.9971 |
| ICEBERG | BIBOP (600) | 0.5200 |
| BURBOP_pm10 | asym. | 0.9979 |
| BURBOP_pm10 | sym. | 0.9981 |
| BURBOP_pm10 | quasi-adiab. | 0.9984 |
| BURBOP_pm10 | BIBOP (600) | 0.5194 |
| BURBOP_pm0 | asym. | 0.9971 |
| BURBOP_pm0 | sym. | 0.9975 |
| BURBOP_pm0 | quasi-adiab. | 0.9976 |

could be adjusted correctly because the R value was known. A better comparison would also include a 1. order phase correction for hard pulses and will be part of further studies as well as the performance of hard pulses as refocusing pulses.

For the PROJECT sequence, three types of pulses were needed: An excitation pulse, a refocusing pulse or a pair of inversion pulses and universal rotation pulses for the *perfect echo* (PE). A 90° UR pulse could also be used for excitation, but not the other way around. Therefore the ICEBERG pulse was only used for excitation.

Again, the BURBOP_pm10 pulse was the best excitation pulse (see Table 5). Depending on the used refocusing pulse(s), $\Phi$ is in range of $0.9979 - 0.9984$. The other BURBOP pulse and the ICEBERG pulse had only slightly lower quality factors. The performance of hard pulses was as bad as in the CPMG sequence. In contrast to the CPMG sequence, using 2 BIBOP pulses for refocusing resulted in a very good performance of the sequence. In combination with the BURBOP_pm10 pulse, the quality factors were significantly higher than for other refocusing pulses. Two 600 µs pulses showed better performance than two 550 µs pulses, probably because the first pulse had a higher quality factor for inversion.

**Simulations including scalar coupling**

With a scalar coupling of $J = 15\,Hz$, which is a good estimation for vicinal F-F couplings [82], the quality factors were very different (see Tables 6 and 7). Only the best excitation pulses have been used in these simulations.

Table 5: Global quality factors Φ of different PROJECT pulse sequences in the absence of J coupling and relaxation

| Excitation pulse | Refocusing pulse | 90° pulse in the PE | Φ |
|---|---|---|---|
| Hard 90°$_x$ | BURBOP_asym | Hard 90° | 0.2362 |
| Hard 90°$_x$ | BURBOP_sym | Hard 90° | 0.2455 |
| ICEBERG | asym. | BURBOP_pm10 | 0.9927 |
| ICEBERG | sym. | BURBOP_pm10 | 0.9940 |
| ICEBERG | quasi-adiab. | BURBOP_pm10 | 0.9945 |
| ICEBERG | 2 × BIBOP (550) | BURBOP_pm10 | 0.9949 |
| ICEBERG | 2 × BIBOP (600) | BURBOP_pm10 | 0.9949 |
| BURBOP_pm10 | asym. | BURBOP_pm10 | 0.9928 |
| BURBOP_pm10 | sym. | BURBOP_pm10 | 0.9937 |
| BURBOP_pm10 | quasi-adiab. | BURBOP_pm10 | 0.9949 |
| BURBOP_pm10 | 2 × BIBOP (550) | BURBOP_pm10 | 0.9966 |
| BURBOP_pm10 | 2 × BIBOP (600) | BURBOP_pm10 | 0.9967 |
| BURBOP_pm0 | asym. | BURBOP_pm0 | 0.9872 |
| BURBOP_pm0 | sym. | BURBOP_pm0 | 0.9932 |
| BURBOP_pm0 | quasi-adiab. | BURBOP_pm0 | 0.9942 |

Most of the CPMG sequences showed very poor performance this time. This was expected because this type of sequence in general cannot refocus homonuclear scalar couplings. When 600 µs BIBOP pulses was used for refocusing, the performance was better than in the absence of scalar coupling: $\Phi = 0.7898$ in combination with the ICEBERG pulse and $\Phi = 0.7905$ in combination with the BURBOP_pm10 pulse. One reason, why these sequences were best CPMG-sequences could be the much shorter duration of the inversion pulses compared to the BURBOP pulses ($t_p = 1$ ms). Sequences using quasi-adiabatic pulses for refocusing had quality factors of $0.5468$ and $0.5471$. Other pulse sequences have quality factors below 20 %.

The performance of the PROJECT sequences was much better: all quality factors were in the range of $0.9869 - 0.9955$. The best refocusing pulses were BIBOP pulses, like in the case without scalar coupling. The decrease in $\Phi$ due to the coupling is less than 1 % in all the cases. An improved performance compared to CPMG is achieved because scalar couplings are refocused through the perfect echoes. The PROJECT experiment has been specifically designed for this purpose.

All in all, it depends on the experimental setup which sequence performs better. In the absence of scalar coupling, the CPMG sequence showed better performance than the PROJECT sequence (see Fig. 30.A). Therefore, in experiments with mono-fluorinated molecules, the CPMG sequence should be the first choice. The best quality factor that could be achieved for the given

**Table 6:** Global quality factors $\Phi$ of different CPMG pulse sequences in the absence of relaxation, J=15 Hz

| Excitation pulse | Refocusing pulse | $\Phi$ |
|---|---|---|
| ICEBERG | asym. | 0.1806 |
| ICEBERG | sym. | 0.0968 |
| ICEBERG | quasi-adiab. | 0.5468 |
| ICEBERG | BIBOP (600) | 0.7898 |
| BURBOP_ pm10 | asym. | 0.1807 |
| BURBOP_ pm10 | sym. | 0.0971 |
| BURBOP_ pm10 | quasi-adiab. | 0.5471 |
| BURBOP_ pm10 | BIBOP (600) | 0.7905 |

**Table 7:** Global quality factors $\Phi$ of different PROJECT pulse sequences in the absence of relaxation, J=15 Hz

| Excitation pulse | Refocusing pulse | 90°pulse in the PE | $\Phi$ |
|---|---|---|---|
| ICEBERG | asym | BURBOP_pm10 | 0.9869 |
| ICEBERG | sym | BURBOP_pm10 | 0.9921 |
| ICEBERG | quasi-adiab. | BURBOP_pm10 | 0.9922 |
| ICEBERG | 2 × BIBOP (550) | BURBOP_pm10 | 0.9936 |
| ICEBERG | 2 × BIBOP (600) | BURBOP_pm10 | 0.9943 |
| BURBOP_pm10 | asym | BURBOP_pm10 | 0.9910 |
| BURBOP_pm10 | sym | BURBOP_pm10 | 0.9931 |
| BURBOP_pm10 | quasi-adiab. | BURBOP_pm10 | 0.9940 |
| BURBOP_pm10 | 2 × BIBOP (550) | BURBOP_pm10 | 0.9955 |
| BURBOP_pm10 | 2 × BIBOP (600) | BURBOP_pm10 | 0.9949 |

parameters was 0.9984. This could be different for other $\Delta$ and $n$ values, which should be investigated in a further study.

In the presence of scalar F-F-couplings, the best achievable quality factor using CPMG was 0.7905. Most offset profiles showed a massive drop in the middle of the offset range, like in Fig. 30. A more uniform profile is necessary for a high accuracy of the experiment. Using the PROJECT experiment led to better results, especially when using pairs of BIBOP pulses for refocusing. The best quality factor was 0.9955 and the offset profiles were very uniform compared to CPMG. However, the performance could be different for other values of experimental parameters. In practice, when working with polyfluorinated organic compounds, scalar couplings can have larger values: Geminal F-F coupling constants can have values up to 300 Hz [82].

The superior performance of the PROJECT sequence has also been shown experimentally by Calle et al. [44], including molecules with a few different geminal and vicinal F-F couplings. In that study, however, only a small part of the fluorine spectrum was investigated and, therefore, hard pulses could be used without problems.

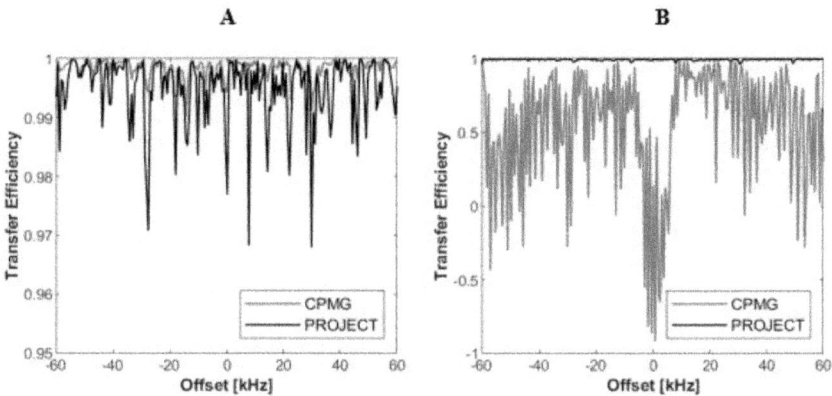

**Figure 30:** In-phase transverse magnetization after applying a CPMG/PROJECT sequence in a system without F-F-coupling (A) and in a system with F-F coupling with $J = 15$ Hz (B). The delay was 2.5 ms in both cases: for CPMG, n=80; for PROJECT, n=40. Thus, both experiments had approx. the same duration. In all cases, BURBOP_pm10 pulse was used as the 90° pulse and the quasi-adiabatic BURBOP pulse was used for refocusing.

# 5   Conclusions and outlook

## 5.1   Broadband 30° and 60° pulses

As an extension to systemetic studies of BEBOP/BIBOP [7], power-BEBOP/power-BIBOP [8], and BURBOP pulses [9], the GRAPE algorithm [1] has been applied to optimize 30° and 60° pulses under various constraints.

Similar to previous studies, rf-power limited excitation pulses showed better performance than rf-amplitude limited pulses, especially for larger bandwidths. In addition, pulse durations of BEBOP and power-BEBOP pulses with different flip angles $\alpha$ have been compared. It was established that the minimum pulse duration for achieving a quality factor of 0.98 increases with respect to the flip angle and the bandwidth. The relation between the minimum pulse duration and $\alpha$ is non-linear. The local quality factors obtained for different frequency offsets showed that rf-amplitude restricted pulses were more selective than corresponding rf-power restricted pulses. The pulse selectivity increases with pulse duration and is higher for BEBOP-60 pulses than for BEBOP-30 pulses.

The observed pulse shapes are similar to previously published BEBOP-90/power-BEBOP-90 pulses. As previously reported for power-BEBOP-90 pulses, the maximum rf-amplitude of rf-power limited pulses is reached in the end of the pulse and its value is the same order of magnitude as the desired bandwidth. This explains their superior performance compared to rf-amplitude restricted pulses.

Different target propagators were used for optimization of BURBOP-30 and BURBOP-60 pulses. The best performance for very short pulse durations is achieved if the target propagator corresponds to a $30°/60°$ rotation instead of a $-330°/300°$ rotation. For long pulse durations, however, both approaches result in comparable quality factors. Rf-amplitude and rf-power limited pulses show similar performance in most of the cases. The minimum pulse durations are significantly larger than for corresponding point-to-point pulses. In addition, BURBOP pulses with different flip angles have been compared. For obtaining a quality factor of 0.99, small flip angle BURBOP pulses have to be almost as long as corresponding BURBOP-90 and BURBOP-180 pulses. This indicates that most of the rf-energy is used for offset compensation and little for the actual flip angle. The shapes of power-BURBOP pulses were symmetric and similar to RE-BURP [81] and ICEBERG pulses [78].

Considering future research, the best point-to-point and universal rotation pulses should be tested experimentally in $\beta$-angle based $^{13}$C experiments as well as two-dimensional heteronuclear experiments. Their performance should be compared to hard pulses and RADFA pulses. Future work could also include optimization of ICEBERG pulses with flip angles of 30° and 60°.

## 5.2   Design of broadband CPMG and PROJECT sequences

The second part of the work was concerned with optimizations of BURBOP-90 and BIBOP pulses for a bandwidth of $120\,\text{kHz}$ and theoretical comparison of the resulting $^{19}$F-CPMG [10, 11] and PROJECT [12] sequences.

When optimizing 90° universal rotation pulses, random pulse shapes as well as CHIRP [53] and BIR-4 [65] shapes were used as starting pulses. For pulse durations close to 2 ms CHIRP-based pulses performed best. However, these pulses were too long for the given application. In other cases, pulses obtained from 100 optimizations with random starting shapes showed best performance.

Multiple approaches were used to design the shortest possible BIBOP pulses with a quality factor close to 0.999. Increasing the number of optimizations from 100 to 500 was not as sufficient as multiple consecutive optimizations with smaller convergence parameters $\delta$. With the latter method, the performance could be improved significantly, but once $\delta = 10^{-8}$, very little improvement was achieved.

The best pulses optimized for the required bandwidth as well as other ultra-broadband pulses available were combined to form CPMG and PROJECT sequences. Simulations showed that CPMG sequences performed slightly better than PROJECT sequences in systems without fluorine-fluorine coupling. In the presence of weak scalar coupling, PROJECT sequences had significantly higher quality factors than CPMG. Consequently, for screening of poly-fluorinated compound libraries it is better to use PROJECT. The best CPMG sequence was a combination of a rf-amplitude variation compensated BURBOP-90 pulse with a BURBOP-180 pulse. For PROJECT, the best results were achieved using the same 90° pulse and a pair of BIBOP pulses instead of a universal rotation pulse.

In future research, simulations with different delays and numbers of echo iterations should be performed because these parameters should be adjustable. In addition, different values of coupling constants should be included in simulations. Here, a value typical for vicinal coupling (15 Hz) was assumed. In general, fluorine-fluorine couplings reach up to $300\,\text{Hz}$ [82]. Therefore the coupling evolution during the sequences should also be examined. Often the performance of repetitive sequences like CPMG can be improved by using phase cycles [7]. Different phase cycles could be implemented in simulations as well as tested experimentally.

# Appendices

## A   Pulses used in the CPMG and PROJECT sequences

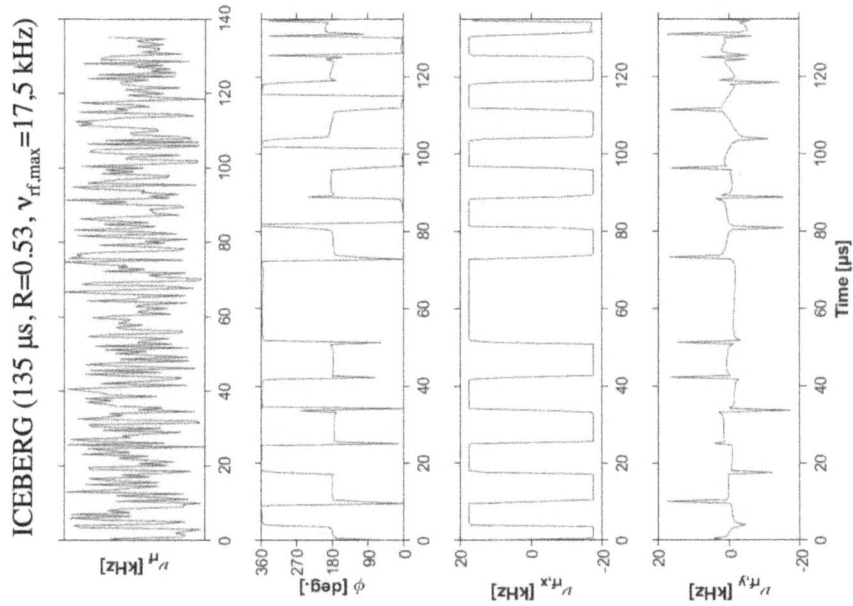

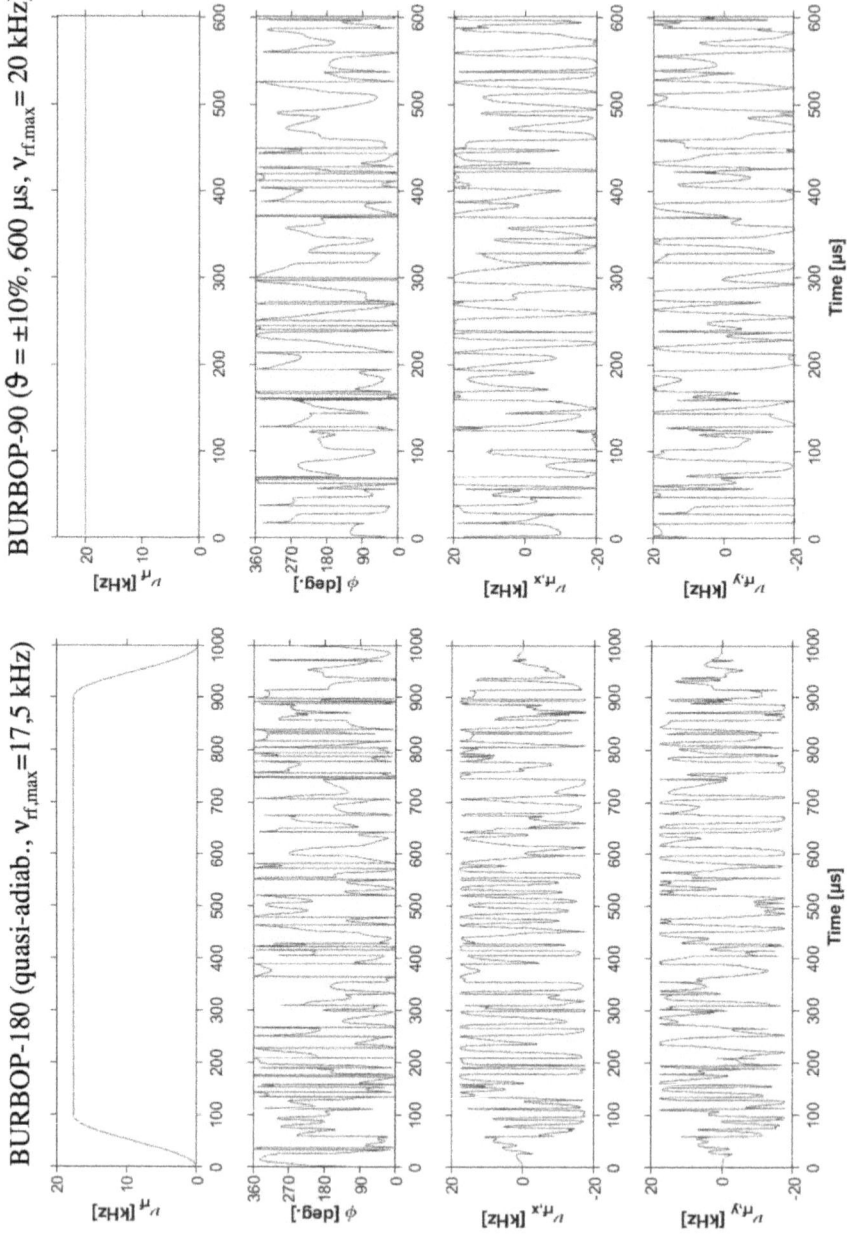

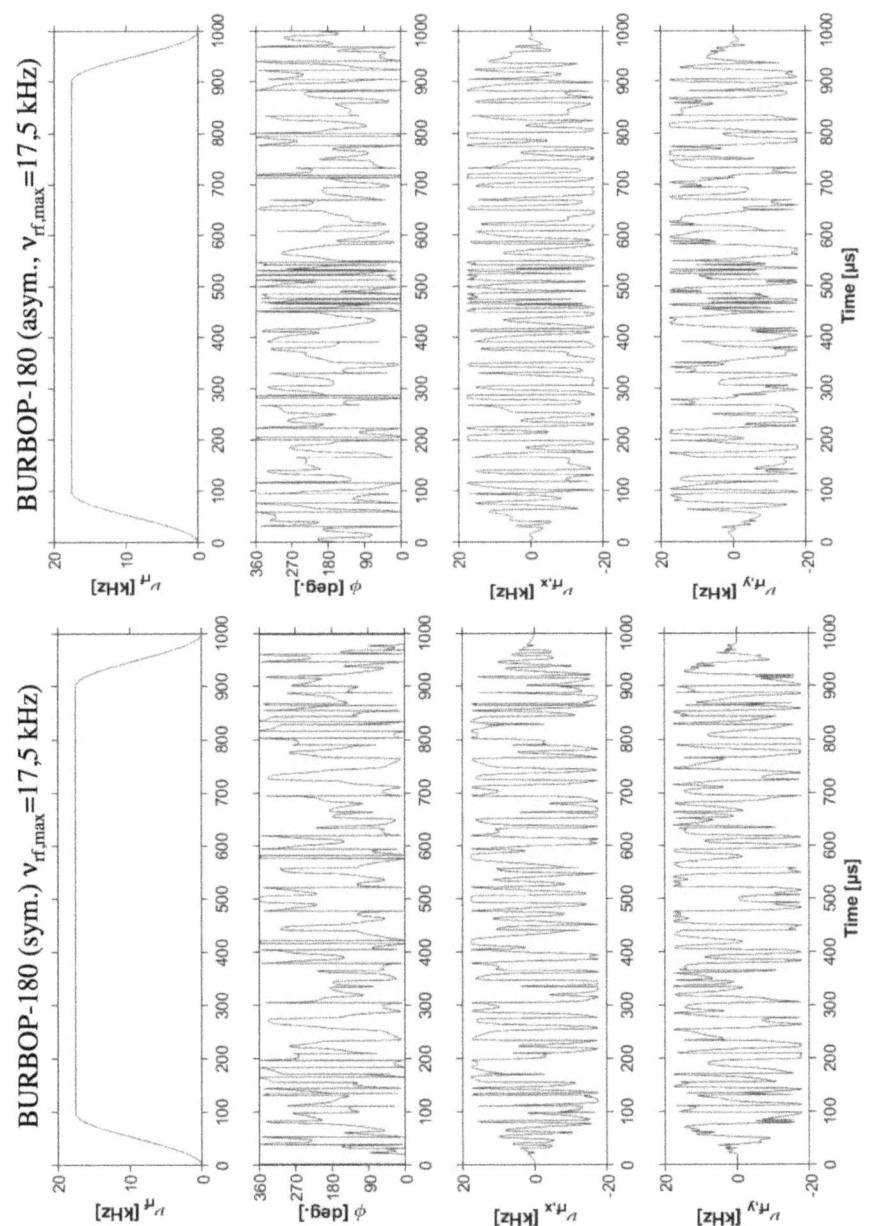

# References

[1] Navin Khaneja, Timo Reiss, Cindie Kehlet, Thomas Schulte-Herbrüggen, and Steffen J. Glaser. Optimal control of coupled spin dynamics: design of NMR pulse sequences by gradient ascent algorithms. *Journal of Magnetic Resonance*, 172(2):296–305, feb 2005.

[2] Steven Conolly, Dwight Nishimura, and Albert Macovski. Optimal Control Solutions to the Magnetic Resonance Selective Excitation Problem. *IEEE Transactions on Medical Imaging*, 5(2):106–115, jun 1986.

[3] Jintong Mao, T. H. Mareci, K. N. Scott, and E. R. Andrew. Selective inversion radiofrequency pulses by optimal control. *Journal of Magnetic Resonance (1969)*, 70(2):310–318, nov 1986.

[4] Daniel Rosenfeld and Yuval Zur. Design of adiabatic selective pulses using optimal control theory. *Magnetic Resonance in Medicine*, 36(3):401–409, sep 1996.

[5] Thomas E. Skinner, Timo O. Reiss, Burkhard Luy, Navin Khaneja, and Steffen J. Glaser. Application of optimal control theory to the design of broadband excitation pulses for high-resolution NMR. *Journal of Magnetic Resonance*, 163(1):8–15, 2003.

[6] Thomas E. Skinner, Timo O. Reiss, Burkhard Luy, Navin Khaneja, and Steffen J. Glaser. Reducing the duration of broadband excitation pulses using optimal control with limited RF amplitude. *Journal of Magnetic Resonance*, 167(1):68–74, mar 2004.

[7] Kyryl Kobzar, Thomas E. Skinner, Navin Khaneja, Steffen J. Glaser, and Burkhard Luy. Exploring the limits of broadband excitation and inversion pulses. *Journal of Magnetic Resonance*, 170(2):236–243, oct 2004.

[8] Kyryl Kobzar, Thomas E. Skinner, Navin Khaneja, Steffen J. Glaser, and Burkhard Luy. Exploring the limits of broadband excitation and inversion: II. Rf-power optimized pulses. *Journal of Magnetic Resonance*, 2008.

[9] Kyryl Kobzar, Sebastian Ehni, Thomas E. Skinner, Steffen J. Glaser, and Burkhard Luy. Exploring the limits of broadband 90° and 180° universal rotation pulses. *Journal of Magnetic Resonance*, 2012.

[10] H. Y. Carr and E. M. Purcell. Effects of Diffusion on Free Precession in Nuclear Magnetic Resonance Experiments. *Physical Review*, 94(3):630–638, may 1954.

[11] S. Meiboom and D. Gill. Modified spin-echo method for measuring nuclear relaxation times. *Review of Scientific Instruments*, 1958.

[12] Juan A. Aguilar, Mathias Nilsson, Geoffrey Bodenhausen, and Gareth A. Morris. Spin echo NMR spectra without J modulation. *Chem. Commun.*, 48(6):811–813, 2012.

[13] Horst Friebolin. *Ein- und zweidimensionale NMR-Spektroskopie.* WILEY-VCH, 5. edition, 2013.

[14] John H.F. Bothwell and Julian L. Griffin. An introduction to biological nuclear magnetic resonance spectroscopy. *Biological Reviews*, 86(2):493–510, 2011.

[15] David S. Wishart. Quantitative metabolomics using NMR. *TrAC - Trends in Analytical Chemistry*, 2008.

[16] Robert Powers. NMR metabolomics and drug discovery. *Magnetic Resonance in Chemistry*, 2009.

[17] T. A. Cross and S. J. Opella. Solid-state NMR structural studies of peptides and proteins in membranes. *Current Opinion in Structural Biology*, 4(4):574–581, 1994.

[18] Alvar D. Gossert and Wolfgang Jahnke. NMR in drug discovery: A practical guide to identification and validation of ligands interacting with biological macromolecules. *Progress in Nuclear Magnetic Resonance Spectroscopy*, 97:82–125, 2016.

[19] Walter Becker, Krishna Chaitanya Bhattiprolu, Nina Gubensäk, and Klaus Zangger. Investigating Protein–Ligand Interactions by Solution Nuclear Magnetic Resonance Spectroscopy. *ChemPhysChem*, 19(8):895–906, 2018.

[20] Friedrich Lottspeich and Joachim W. Engels. *Bioanalytik.* Spektrum Akademischer Verlag, Heidelberg, 3. edition, 2012.

[21] Malcolm H. Levitt and Ray Freeman. NMR population inversion using a composite pulse. *Journal of Magnetic Resonance*, 33(2):473–476, feb 1979.

[22] Malcolm H. Levitt. Composite Pulses. *Progress in Nuclear Magnetic Resonance Spectroscopy*, 18(2):61–122, 1986.

[23] Richard R. Ernst, Geoffrey Bodenhausen, and Alexander Wokaun. *Principles of Nuclear Magnetic Resonance in One and Two Dimensions*. Oxford University Press, 1987.

[24] G. Bodenhausen, G. Wagner, M. Rance, O. W. Sørensen, K. Wüthrich, and R. R. Ernst. Longitudinal two-spin order in 2D exchange spectroscopy (NOESY). *Journal of Magnetic Resonance (1969)*, 59(3):542–550, oct 1984.

[25] Gerhard Wagner, Geoffrey Bodenhausen, Norbert Mueller, Mark Rance, Ole W. Soerensen, Richard R. Ernst, and Kurt Wuethrich. Exchange of two-spin order in nuclear magnetic resonance: separation of exchange and cross-relaxation processes. *Journal of the American Chemical Society*, 107(23):6440–6446, nov 1985.

[26] Erik R. P. Zuiderweg. Complete NMR fingerprints of proteins in $H_2O$ solution without solvent presaturation. An application of two-dimensional longitudinal two-spin-order spectroscopy. *Journal of Magnetic Resonance (1969)*, 71(2):283–293, feb 1987.

[27] W. P. Aue, E. Bartholdi, and R. R. Ernst. Two-dimensional spectroscopy. Application to nuclear magnetic resonance. *The Journal of Chemical Physics*, 1976.

[28] Ad Bax and Ray Freeman. Relative signs of NMR spin coupling constants by two-dimensional Fourier transform spectroscopy. *Journal of Magnetic Resonance (1969)*, 45(1):177–181, oct 1981.

[29] Ad Bax and Ray Freeman. Investigation of complex networks of spin-spin coupling by two-dimensional NMR. *Journal of Magnetic Resonance (1969)*, 44(3):542–561, sep 1981.

[30] Hartmut Oschkinat, Annalisa Pastore, Peter Pfändler, and Geoffrey Bodenhausen. Two-dimensional correlation of directly and remotely connected transitions by z-filtered COSY. *Journal of Magnetic Resonance (1969)*, 69(3):559–566, oct 1986.

[31] Hartmut Oschkinat, Annalisa Pastore, and Geoffrey Bodenhausen. Determination of relaxation pathways in coupled spin systems by two-dimensional NMR exchange spectroscopy with small flip angles. *Journal of the American Chemical Society*, 109(13):4110–4111, jun 1987.

[32] Andrew J. Pell, Richard A. E. Edden, and James Keeler. Broadband proton-decoupled proton spectra. *Magnetic Resonance in Chemistry*, 45(4):296–316, apr 2007.

[33] Pavleta Tzvetkova, Svetlana Simova, and Burkhard Luy. P.E.HSQC: A simple experiment for simultaneous and sign-sensitive measurement of $(^1J_{CH}+D_{CH})$ and $(^2J_{HH}+D_{HH})$ couplings. *Journal of Magnetic Resonance*, 186(2):193–200, jun 2007.

[34] Josep Saurí, Pau Nolis, Laura Castañar, Albert Virgili, and Teodor Parella. P.E.HSQMBC: Simultaneous measurement of proton–proton and proton–carbon coupling constants. *Journal of Magnetic Resonance*, 224:101–106, nov 2012.

[35] Paul Schanda, Eriks Kupče, and Bernhard Brutscher. SOFAST-HMQC Experiments for Recording Two-dimensional Deteronuclear Correlation Spectra of Proteins within a Few Seconds. *Journal of Biomolecular NMR*, 33(4):199–211, dec 2005.

[36] Thomas Kern, Paul Schanda, and Bernhard Brutscher. Sensitivity-enhanced IPAP-SOFAST-HMQC for fast-pulsing 2D NMR with reduced radiofrequency load. *Journal of Magnetic Resonance*, 190(2):333–338, feb 2008.

[37] Bharathwaj Sathyamoorthy, Janghyun Lee, Isaac Kimsey, Laura R. Ganser, and Hashim Al-Hashimi. Development and application of aromatic $[^{13}C, 1H]$ SOFAST-HMQC NMR experiment for nucleic acids. *Journal of Biomolecular NMR*, 60(2-3):77–83, nov 2014.

[38] Martin R. M. Koos, Hannes Feyrer, and Burkhard Luy. Broadband excitation pulses with variable RF amplitude-dependent flip angle (RADFA). *Magnetic Resonance in Chemistry*, 2015.

[39] Martin R. M. Koos, Hannes Feyrer, and Burkhard Luy. Broadband RF-amplitude-dependent flip angle pulses with linear phase slope. *Magnetic Resonance in Chemistry*, 2017.

[40] Claudio Dalvit, Paul E. Fagerness, Daneen T. A. Hadden, Ronald W. Sarver, and Brian J. Stockman. Fluorine-NMR Experiments for High-Throughput Screening: Theoretical Aspects, Practical Considerations, and Range of Applicability. *Journal of the American Chemical Society*, 125(25):7696–7703, jun 2003.

[41] Leszek Poppe, Timothy S. Harvey, Christopher Mohr, James Zondlo, Christopher M. Tegley, Opas Nuanmanee, and Janet Cheetham. Discovery of Ligands for Nurr1 by Combined Use of NMR Screening with Different Isotopic and Spin-Labeling Strategies. *Journal of Biomolecular Screening*, 12(3):301–311, apr 2007.

[42] Claudio Dalvit. NMR methods in fragment screening: theory and a comparison with other biophysical techniques. *Drug Discovery Today*, 14(21-22):1051–1057, nov 2009.

[43] Anna Vulpetti, Ulrich Hommel, Gregory Landrum, Richard Lewis, and Claudio Dalvit. Design and NMR-Based Screening of LEF, a Library of Chemical Fragments with Different Local Environment of Fluorine. *Journal of the American Chemical Society*, 131(36):12949–12959, sep 2009.

[44] Luis Pablo Calle and Juan Félix Espinosa. An improved $^{19}$F-CPMG scheme for detecting binding of polyfluorinated molecules to biological receptors, apr 2017.

[45] James Keeler. *Understanding NMR Spectroscopy*. WILEY, 2. edition, 2010.

[46] Robert F. Stengel. *Optimal control and estimation*. Dover Publications, Inc., 1994.

[47] Alberto Tannús and Michael Garwood. Adiabatic pulses. *NMR in Biomedicine*, 10(8):423–434, 1997.

[48] Michael Garwood and Lance DelaBarre. The Return of the Frequency Sweep: Designing Adiabatic Pulses for Contemporary NMR. *Journal of Magnetic Resonance*, 153(2):155–177, dec 2001.

[49] F. Bloch, W. W. Hansen, and Martin Packard. Nuclear Induction. *Physical Review*, 69(3-4):127–127, feb 1946.

[50] E. M. Purcell, H. C. Torrey, and R. V. Pound. Resonance Absorption by Nuclear Magnetic Moments in a Solid. *Physical Review*, 69(1-2):37–38, jan 1946.

[51] A. Abragam. *The Principles of Nuclear Magnetism*. Oxford University Press, 1961.

[52] M. S. Silver, R. I. Joseph, and D. I. Hoult. Highly selective $\pi/2$ and $\pi$ pulse generation. *Journal of Magnetic Resonance (1969)*, 59(2):347–351, sep 1984.

[53] Jean-Marl Böhlen, Irene Burghardt, Martial Rey, and Geoffrey Bodenhausen. Frequency-modulated "Chirp" pulses for broadband inversion recovery in magnetic resonance. *Journal of Magnetic Resonance (1969)*, 90(1):183–191, oct 1990.

[54] Eriks Kupče and Ray Freeman. Stretched Adiabatic Pulses for Broadband Spin Inversion. *Journal of Magnetic Resonance, Series A*, 117(2):246–256, 1995.

[55] Eriks Kupče and Ray Freeman. Optimized Adiabatic Pulses for Wideband Spin Inversion. *Journal of Magnetic Resonance, Series A*, 118(2):299–303, feb 1996.

[56] Brian D. Ross, Hellmut Merkle, Kristy Hendrich, R. Scott Staewen, and Michael Garwood. Spatially localized in vivo $^1H$ magnetic resonance spectroscopy of an intracerebral rat glioma. *Magnetic Resonance in Medicine*, 23(1):96–108, jan 1992.

[57] Daniel G. Schupp, Hellmut Merkle, Jutta M. Ellermann, Yong Ke, and Michael Garwood. Localized detection of glioma glycolysis using edited 1H MRS. *Magnetic Resonance in Medicine*, 30(1):18–27, 1993.

[58] R. A. de Graaf, Y. Luo, M. Terpstra, H. Merkle, and M. Garwood. A New Localization Method Using an Adiabatic Pulse, BIR-4. *Journal of Magnetic Resonance, Series B*, 106(3):245–252, mar 1995.

[59] Tsang Lin Hwang, Peter C. M. Van Zijl, and Michael Garwood. Asymmetric Adiabatic Pulses for NH Selection. *Journal of Magnetic Resonance*, 138(1):173–177, 1999.

[60] Michael Garwood and Hellmut Merkle. Heteronuclear spectral editing with adiabatic pulses. *Journal of Magnetic Resonance (1969)*, 94(1):180–185, 1991.

[61] R. A. de Graaf, Y. Luo, M. Terpstra, and M. Garwood. Spectral Editing with Adiabatic Pulses. *Journal of Magnetic Resonance, Series B*, 109(2):184–193, 1995.

[62] P. C. M. Van Zijl, T. L. Hwang, M. O'Neil Johnson, and M. Garwood. Optimized excitation and automation for high-resolution NMR using $B_1$-insensitive rotation pulses. *Journal of the American Chemical Society*, 118(23):5510–5511, 1996.

[63] Hellmut Merkle, Haoran Wei, Michael Garwood, and Kâmil Uğurbil. $B_1$-insensitive heteronuclear adiabatic polarization transfer for signal enhancement. *Journal of Magnetic Resonance (1969)*, 99(3):480–494, 1992.

[64] Seong-Gi Kim and Michael Garwood. Double DEPT using adiabatic pulses. Indirect heteronuclear $T_1$ measurement with $B_1$ insensitivity. *Journal of Magnetic Resonance (1969)*, 99(3):660–667, oct 1992.

[65] Michael Garwood and Yong Ke. Symmetric pulses to induce arbitrary flip angles with compensation for rf inhomogeneity and resonance offsets. *Journal of Magnetic Resonance (1969)*, 1991.

[66] R. S. Staewen, A. J. Johnson, B. D. Ross, T. Parrish, H. Merkle, and M. Garwood. 3-D FLASH imaging using a single surface coil and a new adiabatic pulse, BIR-4. *Investigative radiology*, 25(5):559–67, may 1990.

[67] Jean-Marl Böhlen and Geoffrey Bodenhausen. Experimental Aspects of Chirp NMR Spectroscopy. *Journal of Magnetic Resonance, Series A*, 102(3):293–301, may 1993.

[68] Bernd Reif, Matthias Köck, Rainer Kerssebaum, Heonjoong Kang, William Fenical, and Christian Griesinger. ADEQUATE, a New Set of Experiments to Determine the Constitution of Small Molecules at Natural Abundance. *Journal of Magnetic Resonance, Series A*, 118(2):282–285, feb 1996.

[69] Angelo C. Pinto, Susan K. Do Prado, Raimundo Braz Filho, William E. Hull, Andras Neszmelyi, and Gabor Lukacs. Natural abundance $^{13}$C-$^{13}$C coupling constants observed via double quantum coherence. Structural elucidation by the one- and the two-dimensional NMR experiements of velloziolone, a new seco-diterpene. *Tetrahedron Letters*, 23(50):5267–5270, jan 1982.

[70] Jakob Bunkenborg, Niels C. Nielsen, and Ole W. Sørensen. Doubling the sensitivity of natural abundance $^{13}$C-$^{13}$C INADEQUATE with off-resonance compensation. *Magnetic Resonance in Chemistry*, 38(1):58–61, jan 2000.

[71] Lan Jin and Dušan Uhrín. $^{13}$C-detected IPAP-INADEQUATE for simultaneous measurement of one-bond and long-range scalar or residual dipolar coupling constants. *Magnetic Resonance in Chemistry*, 45(8):628–633, aug 2007.

[72] Arthur G. Palmer, John Cavanagh, R. Andrew Byrd, and Mark Rance. Sensitivity improvement in three-dimensional heteronuclear correlation NMR spectroscopy. *Journal of Magnetic Resonance (1969)*, 96(2):416–424, 1992.

[73] Lewis E. Kay, Paul Keifer, and Tim Saarinen. Pure Absorption Gradient Enhanced Heteronuclear Single Quantum Correlation Spectroscopy with Improved Sensitivity. *Journal of the American Chemical Society*, 114(26):10663–10665, 1992.

[74] J. Schleucher, M. Schwendinger, M. Sattler, P. Schmidt, O. Schedletzky, S. J. Glaser, O. W. Sørensen, and C. Griesinger. A general enhancement scheme in heteronuclear multidimensional NMR employing pulsed field gradients. *Journal of biomolecular NMR*, 4(2):301–6, mar 1994.

[75] Sebastian Ehni and Burkhard Luy. Robust INEPT and refocused INEPT transfer with compensation of a wide range of couplings, offsets, and $B_1$-field inhomogeneities (COB3). *Journal of Magnetic Resonance*, 247:111–117, oct 2014.

[76] Alexander A. Marchione and Breanna Conklin. Comparison of the Effects of Different $^{19}$F $\pi$ Pulses on the Sensitivity and Phaseability of the $^{19}$F-$^{13}$C HSQC. *Applied Magnetic Resonance*, 48(5):485–499, may 2017.

[77] Alexander A. Marchione and Elizabeth L. Diaz. Broadband $^{19}$F TOCSY using BURBOP-based spin lock. *Journal of Magnetic Resonance*, 286:143–147, jan 2018.

[78] Naum I. Gershenzon, Thomas E. Skinner, Bernhard Brutscher, Navin Khaneja, Manoj Nimbalkar, Burkhard Luy, and Steffen J. Glaser. Linear phase slope in pulse design: Application to coherence transfer. *Journal of Magnetic Resonance*, 192(2):235–243, jun 2008.

[79] Kyryl Kobzar. *Optimal Control, Partial Alignment and More: The Design Of Novel Tools for NMR Spectroscopy of Small Molecules (PhD Thesis)*. TU München, 2007.

[80] E. Kupce and R. Freeman. Wideband Excitation with Polychromatic Pulses. *Journal of Magnetic Resonance, Series A*, 108(2):268–273, jun 1994.

[81] Helen Geen and Ray Freeman. Band-selective radiofrequency pulses. *Journal of Magnetic Resonance (1969)*, 93(1):93–141, jun 1991.

[82] L. Petrakis and C. H. Sederholm. NMR fluorine-fluorine coupling constants in saturated organic compounds. *The Journal of Chemical Physics*, 1961.

# YOUR KNOWLEDGE HAS VALUE

Ingram Content Group UK Ltd.
Milton Keynes UK
UKHW010655050623
422889UK00005B/720